W9-CWM-330

Fundamentals of Engineering Design

W. P. Lewis

A. E. Samuel

Department of Mechanical and Manufacturing Engineering
University of Melbourne

PRENTICE HALL

New York London Toronto Sydney Tokyo

Prentice Hall, Inc., Englewood Cliffs, New Jersey
Prentice Hall of Australia Pty Ltd, Sydney
Prentice Hall Canada, Inc., Toronto
Prentice Hall Hispanoamericana, SA, Mexico
Prentice Hall of India Private Ltd, New Delhi
Prentice Hall International, Inc., London
Prentice Hall of Japan, Inc., Tokyo
Prentice Hall of Southeast Asia Pty Ltd, Singapore
Editora Prentice Hall do Brasil Ltda, Rio de Janeiro

Typeset by: Double Click, Brookvale, NSW
Printed and bound in Australia by:
Impact Printing, Brunswick, Vic.

Cover design by Kim Webber

1 2 3 4 5 93 92 91 90 89
ISBN 0 7248 0476 5 (paperback)
ISBN 0-13-33810-5 (hardback)

National Library of Australia
Cataloguing-in-Publication Data

Lewis, W. P.
Fundamentals of engineering design : ideas, methods and applications

Includes index.
ISBN 0 7248 0476 5.

1. Engineering design. I. Samuel, A.E.
II. Title.

620'.00425

Library of Congress
Cataloguing-in-Publication Data

Lewis, Bill, 1934-

Fundamentals of engineering design : ideas, methods, and application / Bill Lewis and Andrew Samuel.
p. cm.
Bibliography: p.
ISBN 0-13-338310-5
1. Engineering design. I. Samuel, Andrew, 1931-
II. Title.
TA174.S257 1988
620'. 00425--dc 19

88-21916
CTP

PRENTICE HALL

A division of Simon & Schuster

Contents

Preface xi

Introduction xv

1 Theory 1

1.1 What is design? 1
1.2 The concept of 'problem' 2
1.3 The design process 4
1.3.1 Evolution of design problems 4
1.3.2 Operational model of the design process 9
1.3.3 Flow chart 11
1.4 The design hierarchy 11
1.5 Review of design thinking 14
1.5.1 Design thinking and everyday thinking 14
1.5.2 The dilemmas of design 16
1.6 Some important issues 16

2 The divergent phase of the design process 18

2.1 Recognition of design problems 18
2.2 Definition of design problems 19
2.2.1 General 19
2.2.2 Checklists of objectives 19
2.2.3 Design objectives, criteria, and constraints 21
2.2.4 Input-output analysis 22
2.2.5 Matching 24
2.2.6 The initial appreciation 24
2.3 The search for alternative solutions 26
2.3.1 The engineering repertoire 26
2.3.2 Creation of new proposals 27
2.3.3 Aids to creative effort 30
2.3.4 Idea logs 36
2.4 Conclusion 37
2.5 Notes from a designer's workbook 37
2.5.1 The 'bug' list 37
2.5.2 An idea log 37

3 The convergent phase of the design process: Decision making 41

3.1 Introduction 41
3.2 Feasibility studies 42
3.3 Selecting one proposal from a number of alternatives 42
3.3.1 General 42
3.3.2 Probabilities 43

3.4 Strategies of decision making 45
3.4.1 Interacting decisions 45
3.4.2 Sequences of decisions in innovative design 45
3.4.3 Sequences of decisions in evolutionary design 47
3.4.4 Computational strategies and mathematical modelling 51
3.4.5 The major aims of a decision strategy 52
3.5 Compromises 52
3.5.1 Introduction 52
3.5.2 Methods of compromise 54
3.6 Conclusion 57

4 Communicating formal messages: The design report 58

4.1 Introduction 58
4.2 Some common errors and misconceptions 58
4.3 The ingredients of good report writing 59
4.4 A recipe for success 60
4.5 On the lighter side 61

5 Economic factors in design 65

5.1 Introduction 65
5.2 Cost to manufacture 65
5.3 Economic evaluation of projects 67
5.4 Life cycle costs 68

6 Ergonomics 71

6.1 Introduction 71
6.2 People in engineering systems 73
6.2.1 Occupant of workspace 73
6.2.2 Source of power 73
6.2.3 Sensor or transducer 73
6.2.4 Processor of information 74
6.2.5 Tracker and controller 75
6.3 Personal factors 75
6.3.1 General 75
6.3.2 Motivation 76
6.3.3 Aesthetics 76
6.4 Environmental factors 77
6.5 Safety 77
6.5.1 Theory 77
6.5.2 The cost of safe design 78
6.6 Design for handicapped people 78
6.7 Notes from a designer's workbook 79
6.7.1 An example of ergonomics in engineering design 79
6.8 Anthropometric data 84

7 Implementation of design 88

7.1 Introduction 88
7.2 Tolerances in the detailed design of components 89
7.3 Risk analysis: Uncertainties in design variables 91
7.4 Standard codes and specifications 92
7.5 Interaction of design with manufacturing and construction 93
7.6 Notes from a designer's workbook 94
7.6.1 Design for variability in component performance—electronic circuit 94
7.6.2 Design for variability in component geometry—gearbox 96
7.7 An exercise in design for manufacturing variability 98
7.7.1 Design specification for relay contact assembly 98

8 Design for integrity 100

8.1 Design procedures 100
8.1.1 Structural integrity: Design against failure 100
8.1.2 A note on design for failure 101
8.2 Definition of failure 102
8.2.1 General 102
8.2.2 Examples of failure 102
8.3 Prediction of failure 103
8.4 Factors of safety 104
8.4.1 Introduction 104
8.4.2 Implicit factors of safety 105
8.4.3 Explicit factors of safety 105
8.4.4 Probabilities 106
8.5 Design methods 107
8.5.1 Design from first principles: I.D.E.A.S. 107
8.5.2 Design from precedent 108
8.6 Notes from a designer's workbook 109
8.6.1 Design of a tensile testing gripper for concrete test specimens 109
8.6.2 A laboratory chimney 114
8.7 Exercises in design for structural integrity 118
8.7.1 Simple structure 118
8.7.2 Mild steel welded bracket 119
8.7.3 Oil-gas separator 119
8.7.4 Timber stacking platform 119
8.7.5 Hydraulic pit-prop for mines 119
8.7.6 Modular guard rail 122

9 Selection of engineering materials 124

9.1 Overview 124
9.2 Materials 125
9.3 Iron and steel 126
9.3.1 Cast iron 126

9.3.2 Plain carbon steel 128
9.3.3 Alloying steels 132
9.3.4 High-strength low-alloy (HSLA) steels 132
9.3.5 Stainless steel 133
9.3.6 Special steels 133
9.4 Non-ferrous metals 134
9.4.1 Aluminum 134
9.4.2 Copper 134
9.4.3 Nickel 135
9.5 Plastics and other composites 135
9.6 Wood and concrete 136
9.7 Material selection 137
9.8 Notes from a designer's workbook 140
9.8.1 Design of a leaf spring for an overspin clutch 140
9.9 Exercises in material selection 142
9.9.1 Flywheel 142
9.9.2 Pressure vessels 142
9.9.3 Ocean going LPG tanker 142
9.9.4 SCUBA tank 142

10 Fatigue failure and stress concentrations 144

10.1 Dynamic loading: Fatigue 144
10.1.1 Introduction 144
10.1.2 Design of elements subject to reversed axial stress 147
10.1.3 Design of elements subject to fluctuating axial stress 150
10.2 Stress concentrations 151
10.2.1 Change of shape 151
10.2.2 Residual stresses 152
10.3 Notes from a designer's workbook 153
10.3.1 Design of a jack-hammer 153
10.4 Exercises in design for dynamic loads 159
10.4.1 Column for riveting machine 159
10.4.2 Link 160
10.4.3 Bell-crank 160
10.4.4 Torsion bar 161

11 Elements subject to axial, transverse and torsional loads 163

11.1 Elements in tension 163
11.2 Elements in compression 164
11.2.1 Long (or slender) columns 164
11.2.2 Short columns 165
11.2.3 Application to design 166
11.3 Elements subject to combined compression and bending 168
11.4 Elements subject to transverse loads—beams 168
11.5 Elements subject to bending and torsion—shafts 169

11.5.1 Modes of failure 169
11.5.2 Prediction of failure by fatigue 169
11.5.3 Application to design 171
11.6 Notes from a designer's workbook 172
11.6.1 Design of a simple truss 172
11.6.2 Machine shaft design 175
11.7 Exercises in design for axial load, bending and torsion 178
11.7.1 Mobile crane 178
11.7.2 Cold water return pipe 179
11.7.3 Scissor jack failure 180
11.7.4 Playground maypole 181
11.7.5 Industrial fuel tank 182
11.7.6 Tubular column design 183
11.7.7 Hydraulic jack 184
11.7.8 Centrifugal pump shaft 185
11.7.9 Vertical hollow-shaft motor 185
11.7.10 Industrial centrifuge 188
11.7.11 Boom gate 189

12 Elements subject to internal fluid pressure—Pressure vessels 190

12.1 Introduction 190
12.2 Design of the cylindrical shell 192
12.3 Other design details 193
12.4 Other clauses in pressure vessel codes 193
12.5 Notes from a designer's workbook 194
12.5.1 Industrial centrifuge 194
12.6 Exercises in design to resist internal pressure 199
12.6.1 Horizontal pressure vessel 199
12.6.2 Hydraulic ram 200

13 Joints for engineering components 202

13.1 Bolted joints 202
13.2 Bolting 205
13.3 Comments on bolted joints 205
13.4 Pinned joints 208
13.5 Bolted joints—bolts in shear 209
13.6 High tensile steel bolted joints 210
13.7 Introduction to welded joints 210
13.7.1 Butt welds 211
13.7.2 Fillet welds—parallel and transverse loads 211
13.7.3 Fillet welds in bending 213
13.7.4 Design of fillet welds based on allowable load per unit length of weld 214
13.7.5 Fillet welds with bending and shear 215
13.7.6 Fillet welds in torsion 215
13.8 Adhesives 216

13.9 Notes from a designer's workbook 217
13.9.1 Design of a bolted joint for a jack-hammer 217
13.10 Exercises in the design of joints 220
13.10.1 Bolted joint for pressure vessel 220
13.10.2 Cylinder-head studs 221
13.10.3 Hydraulic actuator 222
13.10.4 Flanged coupling 223
13.10.5 Welded bracket 224
13.10.6 Pipe support bracket 224
13.10.7 Roof truss support bracket 224

14 Contactual elements 227

14.1 Introduction 227
14.2 Spherical bodies in contact 228
14.3 Cylindrical bodies in contact 229
14.4 Cyclic loading 230
14.5 Application to gear design 230

Tables of data 231
1. Bending moments and deflections of beams 231
2. Conversion tables 235

Design workbook 236
1. Theory of design 236
2. Problem definition 237
3. Generation of ideas 239
4. Decision making 239
5. Economic evaluation 240
6. Benefit—cost analysis 242
7. Ergonomics 242
8. Safety and failure analysis 243

References 245

Index 251

Preface

Other men look at things that are and ask "why?" - I dream of things that never were and ask "why - not ? "

J. F. Kennedy paraphrasing Theodor von Karman comparing scientists, who ask *"why?"*, to engineers, who ask *"why not?"*

This book describes the theory and practice of engineering design. Throughout, the authors' aims have been:

1. to present and elucidate existing knowledge;
2. to stimulate discussion in seminars and tutorials; and
3. to provide sufficient references from the literature for readers with particular interests to delve deeper into the topics concerned. This text is neither a handbook nor a Bible. Since the science and art of engineering design are continually evolving, it is not possible for the final word to be written on any aspect of it.

There is a Talmudic parable about a man who runs down the street shouting:

I have an answer . . . now what is the question?

This parable highlights the contrast between the results of engineering science and the challenge of applying them in engineering design. In design one must *create answers to new* problems rather than *questions to match existing solutions of old problems*. The process of creation is at once ego building and demoralizing for the young designer. It is ego building, because we create something from our own personal store of skills and knowledge, bringing to 'life' as it were our own personal involvement in a problem and the process of finding a solution to it. It is demoralizing because of its complexity and the considerable degree of apparent arbitrariness involved.

In engineering design, the outcomes of our plans are never precisely determined in advance and we never hope to find unique solutions to our problems. No indeed! The safe and comforting uniqueness of well defined problems with deterministic or closed form solutions are the domain of pure mathematics and the physical sciences.

Engineering design is full of uncertainties. In fact many of the decisions made in developing a design appear to be arbitrary. Often designers are advised to choose a 'first' trial in a design iteration arbitrarily by 'guessing'. In time to come, with experience, we hope that these 'first guesses' will in fact be well informed or 'educated guesses'. These educated guesses are the rules of thumb or 'heuristics' of the experienced engineer. To the young and inexperienced designer, these apparently arbitrary decisions are both unnerving and quite contrary to the discipline of the physical sciences . One often finds young designers asking 'How can we be expected to make such and such a decision?' The usual answer is that 'Someone has to, so why not you ?'

In time, one hopes to gain experience and technical maturity so that these decisions, while nonetheless arbitrary, are certainly less frightening. This book reduces the fear of apparently arbitrary decision making by exposing the student designer to problems with a real-life flavour. Some experienced designers argue that it is difficult to encapsulate real-life experiences in an academic environment. The book invokes real-life problem situations and follow up discussions and coaching sessions must provide the necessary reinforcement in any proper undergraduate design course. The text on its own is not a handbook. It is intended to be a teaching text. The book combines theory with real life design problems and examples of professional engineering solutions. It presents recognizable strategies for solving engineering design problems — 'skyhooks for the mind'.

The book is a result of the authors' 20 years of experience in teaching and educational research in engineering design. Successful design requires a well-balanced mix of the general and the specific, a breadth of vision and sensitivity to the manifold of consequences which may follow some chosen course(s) of action. Yet good designers are unremitting in their attention to detail. Thus the first part of the book provides a broad vision of engineering design and a theoretical and conceptual framework.The second part deals with a more specific approach, involving detailed elements of the designer's art. In the first part a strategic approach is provided for grappling with open-ended loosely defined problems, while in the second part the tactics of solving smaller scale problems are explored. These smaller-scale problems have been 'distilled' from the strategic approach to the larger-scale problems. Strategy and tactics are the essential threads of the large tapestry woven by the competent designer's art.

In order to clarify for the reader the various elements in the book, a standardized format is used in the headings of chapters. In each case we tell the reader what *concepts* are to be explored in the chapter, together with the *skills* to be developed and, where appropriate, examples of *specialized engineering hardware* included in discussions.

The first part of the book comprising Chapters 1 to 7 deals with the process of engineering design, problem finding and problem development. There are examples of the way engineers deal with problems. Dealing with a problem is not necessarily finding a solution to it ab initio. Boguslaw, in his book *The New Utopians*, suggests that the designer is a magician with a wand that has a *finite time constant*. In this sense the 'solution' could involve the removal of the need for a solution.

The objectives of this part of the book are:

1. to introduce the formal procedures of engineering design, and to use these procedures in exploring some typical engineering problems (these matters are dealt with in Chapter 1);
2. to introduce techniques for dealing with large scale open ended problems, where the designer must first gather information and process this for developing a suitable solution path (this 'divergent' phase of the design process is discussed in Chapter 2);
3. to introduce formal procedures to aid in choosing between competing alternatives in design (this is the 'convergent' phase of design and it is discussed in Chapter 3);
4. to review the methods used in presenting the findings of engineering investigations in a formal Design Report (the subject of Chapter 4);
5. to examine the effect of economic and human interactions with our designs (Chapters 5 and 6 deal with these matters);
6. to examine the way our design might interact with a less than perfect world where our ideas must be implemented and where failure (perish the thought) is a possibility (Chapter 7 deals with these matters).

The second part of the book comprising Chapters 8 to 14 is concerned with the design of specific engineering components. Although the examples are drawn mainly from mechanical and civil engineering, the more general procedures of the engineering design process pervade this part of the book also. The accent is on problem finding and problem exploration, even to the extent of evaluating competing alternatives. In this sense the book is unique in its treatment of engineering design.

The objectives of this part of the book are:

1. to impart knowledge of important concepts and methods;
2. to develop appropriate and relevant problem solving skills;
3. to develop professional and responsible attitudes in applying the above.

The final outcome is a capacity for the young engineer to synthesize solutions to practical design problems.

Chapter 8 deals with concepts of failure of engineering structures and components, together with factors of safety. Chapter 9 is a short overview of available engineering materials and some of their performance criteria. Chapters 10 and 11 examine structures and components subjected to a variety of loading conditions. Chapters 12 through 14 consider some special, but nonetheless important, engineering components and their design for structural integrity.

Finally, there are copious examples of solved and unsolved problems throughout the book as well as a workbook of examples of appropriate engineering design problems for the educator. Notably these examples are formatted to encapsulate the character of problems in real-life engineering design as experienced by the authors in their own professional and consulting experiences. All the problems leave some opportunities for the designer to exercise creative talent. None have unique solutions. This is the stuff of engineering design.

WPL/AES
Melbourne, 1988

Introduction

Engineering designers are imaginative, realistic and optimistic.

Imaginative because they are constantly carrying out thought experiments, imagining in their minds how the structures and machines they are designing will perform when built and put into operation. *What if*, they ask themselves, what if these components were longer, shorter, thicker, thinner, in titanium or fiber reinforced plastic instead of steel, and so on, continually exploring alternatives in their attempts to give the client a solution which makes the best use of the resources available.

Einstein was a famous exponent of the thought experiment. His imagining of what would happen if he were able to sit on a light ray and travel with the speed of light — would time stand still? — helped him to the formulation of a four-dimensional space-time continuum for the theory of relativity. Another question he asked:

> *Would an observer stationary at a point on the earth's surface see a distant flash of lightning at the same instant as a second observer travelling in a train towards the place where the flash occurred?*
>
> (Einstein, 1936; Einstein and Infeld, 1938)

It is said that Sir Henry Royce (of Rolls-Royce fame) could run a new engine in his mind, and in so doing his imagination would identify potential weaknesses and regions of wear.

Although these examples are merely a few representatives from a very large pool of experience, the essential point remains. Successful engineering design demands a high level of conceptual thinking. *Rote learning of techniques of mathematical manipulation is anathema.*

Here are some '*What if*'? questions for you, the reader, to think about:

- What would happen if a candle were lit in a spacecraft?
  How would the 'flame' appear?
  Indeed, would there be a flame?
- What would happen if a petrol tanker were to collide with a liquid oxygen tanker at a busy intersection?
  Consider a sphere of any radius (say the radius of the earth). When a central circular cylindrical hole is drilled completely through this sphere, a solid of revolution of length L remains.
- What is the volume of this remaining solid?
  (*Hint*: this volume is a function of L only)
- What if all lecture theatres were painted pale blue? red?
- What if pay tolls were introduced on urban arterial roads?
- What design problems would immediately arise if a law were passed that all central business districts were to be '*no passenger*' car zones.

Successful designers are *realists*. Thought experiments are cheap, the only resource expended is the designer's time. But is this resource sufficient for the task at hand?

The Tacoma Narrows suspension bridge in the United States was constructed according to the best design principles of the time (1940), but failed spectacularly after a few months' service as a result of severe wind-induced vibrations. In a 65 km/hr cross wind large swirling air vortices were shed alternately from sharp corners at the top and bottom of the roadway, acting as a bluff body in a turbulent air stream. The frequency of the vortex shedding and of the associated dynamic forces coincided with a natural frequency of torsional oscillation of the bridge structure (Pugsley, 1968). This was a phenomenon not envisaged by the designers — it was outside their previous experience and foreign to their conceptual thinking about problems in structural design.

In 1857 the steamship the *Great Eastern*, designed by the great Victorian engineer, Isambard Kingdom Brunel, was launched. Approximately 210 meters long and 31,500 tonnes displacement, powered by the largest marine steam engines of the time, and aimed at capturing lucrative transatlantic trade, it was a total failure. It could travel at only one-third of its designed forward speed due to Brunel's quite unrealistic predictions of the propulsive power required by large ships to overcome wave making drag (Rolt, 1957).

Inexorably, the laws of nature will take their toll of engineers who flout them or ignore them.

Designers are *optimists*. The evolving nature of design problems leads them through cycles of success and failure. But at any instant the fear of failure, a very human attribute, is held in check by a balanced and basically optimistic outlook on life — positive yet objective.

Early in 1921 Sir John Monash, a distinguished civil engineer and soldier, was appointed chairman and chief executive officer of the newly formed State Electricity Commission of Victoria, Australia (S.E.C.). The S.E.C. had been set up to exploit enormous brown coal deposits in the Latrobe Valley. By late 1921 and early 1922 the whole enterprise was in crisis. Early mineral exploration and shaft sinkings had indicated the brown coal to have a moisture content of 45 percent to 48 percent. Orders for boilers, supporting structures and associated equipment were placed on this basis. However, continued testing revealed moisture contents 65 percent to 68 percent, unprecedentedly high values for which existing boiler technology was quite unable to cope.

In public and before State Parliament Monash brushed aside, even ignored, the problem until, after two uncertain years of very intensive research and development within the S.E.C., ways were found of redesigning and modifying the boilers and for drying the coal on the way to the combustion zones. Monash and his engineers *were running great risks which required steady nerves: if the experiments to adapt failed, the delay and costs would be incalculable and their secrecy indefensible* (Serle, 1982).

So we have a vision of the designer as *hero* — a person of *self-awareness* and *self-control, imaginative, realistic* and *optimistic* — as the occasion demands.

A university course in engineering design has two major objectives which are:

1. to instruct young engineers in the methods and techniques of engineering design. To this end, general methods of solving design problems and specific techniques for particular applications are described and explained. By working through a graduated program of design exercises, the ability to use these methods and techniques is developed, and a firm foundation laid for future engineering action; and
2. to develop young engineers' ability to combine and apply their knowledge of the

engineering sciences. This knowledge has in turn to be integrated with their own observations and experiences in the complex situations typical of engineering.

If these objectives are attained, young engineers successfully develop their capacity for professional problem solving — and incidentally for problem finding.

What are the attributes of good designers?

First and foremost, they are equally at home with the general and abstract as with the specific and concrete. They are masters of the theory of science and the art of engineering. Doctors can bury their mistakes but designers' errors are exposed for all to see. The avoidance of error requires unremitting attention to detail, to the checking of each link in the chain of decisions — all at the cost of much intellectual 'blood, toil, tears and sweat'.

Designers dream dreams, they see not only the way things are but also the way things ought to be. Designers have the drive and motivation to create new ideas and then the courage to evaluate their creations critically and objectively. The generation and evolution of ideas demands an emotional commitment, but the intellectual discipline of their profession ensures that designers keep their emotions under control.

The vision of great designers encompasses people as well as things; their hopes and aspirations match the needs of the society they serve.

As well as these attributes designers have *knowledge*, *skills* and some *attitudes* commensurate with those for whom the design solution must work, as well as attitudes associated with professional ethics. This includes the knowledge of concepts, such as failure and strength for example. Knowledge of methods includes general design procedures as well as the detailed and specific methods of the relevant engineering sciences, such as drawing free body diagrams for example. The competent designer must have a wide range of skills as listed below:

- Design skills in *applying general methods* and in *organizing work* in accordance with these methods.
- Skill in *foreseeing modes of failure.*
- Skill in *design predictions.*
- Skill in *identifying sources of uncertainty* in design predictions.

  Note: *Judgment in making quantitative allowance for these uncertainties is developed largely by experience. We can, however, make explicit the methods used by professional engineers to cope with uncertainties.*

- Skill in *solving multi-variable problems.* Multi-variable problems arise in engineering designs which are not of a neat mathematical form. Frequently the number of unknowns exceeds the number of equations relating them; there are inequalities to be satisfied; some variables are discontinuous. An important question is *what is the most efficient strategy for exploring this situation*?
- Skill in *appraising results.* The results of design calculations have to be related to the real world. Questions to be asked are: Are the results of the right order of magnitude? To what accuracy should they be quoted? What accuracy is justified bearing in mind possible inaccuracies in the data used and practical difficulties of construction and manufacture?
- Skill in *trial-and-error calculations.* In most cases, direct design or synthesis is not possible. A trial design has to be 'guesstimated', analyzed, and then revised if the analysis reveals faults. The designer has to be ready to try something and 'give it a go'.

- Skill in *making approximations*, taking legitimate shortcuts in what would otherwise be unnecessarily involved calculations. This is not a purely computational skill, but depends also on insight into the physical situation being represented mathematically. It is allied to the ability to recognize key issues and concentrate on them.
- Skill in *assigning tolerances to design variables*. This depends on skill in *analogue simulation* in determining the *sensitivity of the design to change*.
- Skill in *assigning tolerances* to linear dimensions and cylindrical surfaces. This depends on skill in *constructing the path equation*, in selecting the appropriate datum for measurement of dimensions, and in avoiding conflicting tolerances.

Competent engineering designers are able to take into account a great variety of problems. Some of these problems may appear, on the surface, to be quite outside the domain of engineering. Yet challenges to the designers' skills are found everywhere and they often pervade our daily lives. Some sample challenges are:

- Why do sausage skins split parallel to the long axis when barbecued?
- Why is corrugated iron corrugated?
- What is the purpose of an outrigger on a sailing canoe?
- Why do yachtsmen occasionally hang off the side of a yacht on a trapeze when racing?
- Why does a piece of chalk fail along a 45 degree helix when twisted in pure torsion?
- Why is the Eiffel Tower such a distinctive shape?

The real challenge of engineering design is to look at things and see their essential engineering qualities and content. In the words of Jonathan Swift *'There is none so blind as they that will not see'*.

Note: *Our work in engineering design assumes that certain basic knowledge and skills have already been acquired, namely:*

- knowledge of *elementary statics*, *mechanics of solids* and *properties of engineering materials*; and
- skill in *reading and interpreting drawings* and *visualizing in three dimensions*, skill in the construction of *free body diagrams* showing forces in equilibrium, skill in routine mathematical manipulation and calculation.

I know why there are so many people who love chopping wood. In this activity one immediately sees results.

Albert Einstein

Chapter 1

Theory

Like all other sciences, that of the intellectual operations must rest primarily upon observation, the subject of such observation being the very operations and processes of which we desire to determine the laws.

George Boole

Concepts introduced — problem; rational problem-solving behaviour.

Methods presented — the design process; the evolution of a problem.

Applications
(a) accidents at a level crossing;
(b) design of a small, 12 volt, battery powered water heater — a preliminary study.

1.1 What Is design?

Engineering design is a complex problem-solving activity. In essence it comprises the planning of engineering systems, devices, products and components in order to satisfy some human need. To answer the question, '*What is design*?' we shall have to examine all the different kinds of activities it encompasses. Our approach is essentially a behavioural one, set firmly in the context of professional engineering endeavours.

When we observe practicing professional engineers, we find it is possible to describe and classify their behaviour, as shown in Figure 1.1. The initial distinction between cognitive (intellectual) and affective (emotional) behaviour follows that of Bloom et al. (1956) and Kratwohl et al. (1964). Vickers (1965) has argued that a large proportion of professional activity in day-to-day affairs is devoted to the maintenance of ongoing sets of relationships within a human and physical environment.

Our concern, however, is with design as a form of engineering problem solving. Figure 1.1 indicates that there are three major types of engineering problems, distinguished by their different goals, namely:

1. *research* — directed towards an understanding of natural phenomena;
2. *design* — directed towards satisfying some human want or need; and
3. *operations* — directed towards maintaining the effective operation of an engineering system.

Over the years there has been much interest in the methods used by engineers to solve these different types of problem. Research is characterized by the '*Scientific*

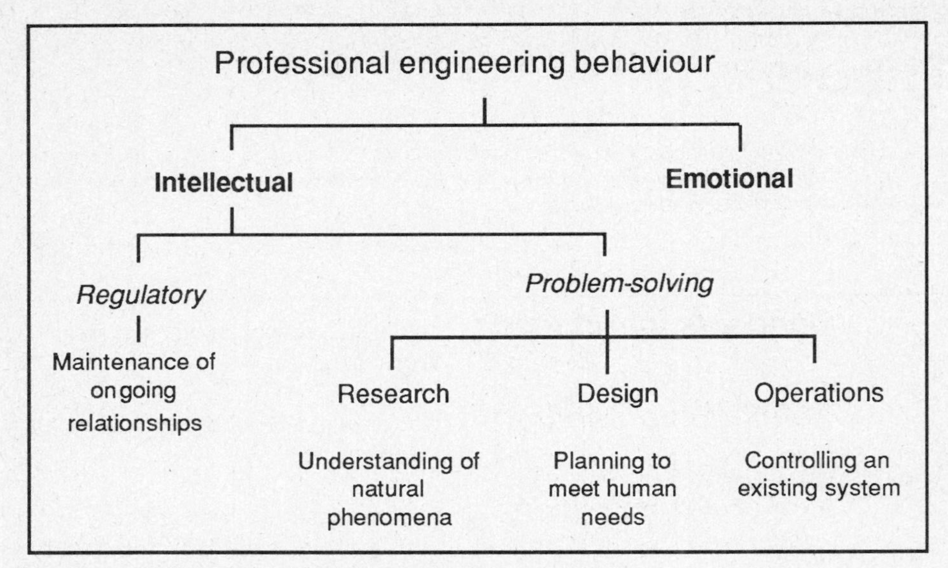

Figure 1.1 Types of professional engineering behaviour

method' (Kuhn, 1970). Design methods have been investigated by Alexander (1964), Alexander and Manheim (1965), Archer (1965, 1969), Asimow (1962), Hall (1962), Hill (1970), Jones (1970), Krick (1969), Luckman (1967), Nadler (1967) and Sandor (1964). We should also note Kepner and Tregoe's formulation of a general method of solving operational problems. Before examining the process of engineering design in detail, we take a philosophical side-glance at the concept of '*problem*'.

1.2 The concept of 'problem'

'*Problem*' is one of those basic and all-embracing words whose meaning we are prepared to accept without close examination. But there are many varieties of problem, as will become evident during the study of engineering design. Some are of interest to engineers, others are not.

We take as the starting point for our discussion the cybernetic model of rational, human, goal-seeking behaviour shown in Figure 1.2 (Heywood, 1970). Problems do not exist in a vacuum, but arise when people — problem solvers — perceive a goal but not the means of attaining it (Gagné, 1959; Merrifield et al., 1962; Krick, 1969).

There is a mismatch between their expectations and the sensory data they are receiving from their environment (and the patterns they recognize in the data). A possible model of the potential problem-solver is shown in Figure 1.3.

George (1970) defines a problem in behavioural terms as a situation to which a human being has to respond (or give a set of responses) but where the response (or set of responses) has not already been acquired by the person concerned. Problem solvers therefore make plans and act in accordance with these plans in a way which they predict will achieve the desired goal. They observe the outcomes of their actions and if there is a difference between the observed outcomes and the desired goal they take corrective action. In attempting to achieve the goal and to maintain conditions at the desired level of performance, problem solvers undertake two major types of activity:

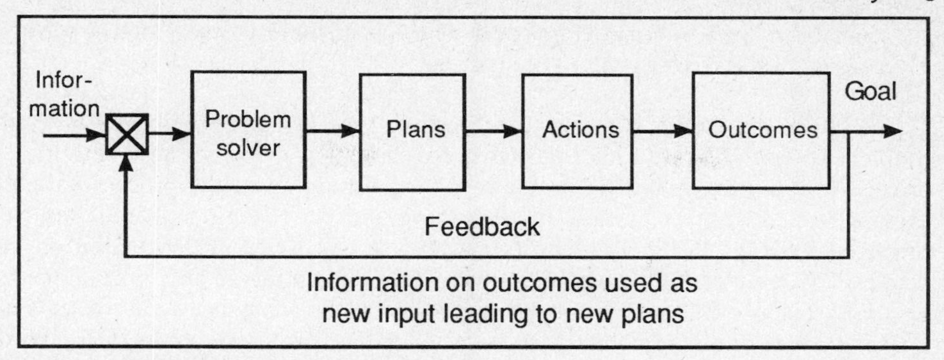

Figure 1.2 Model of rational, goal-seeking behaviour

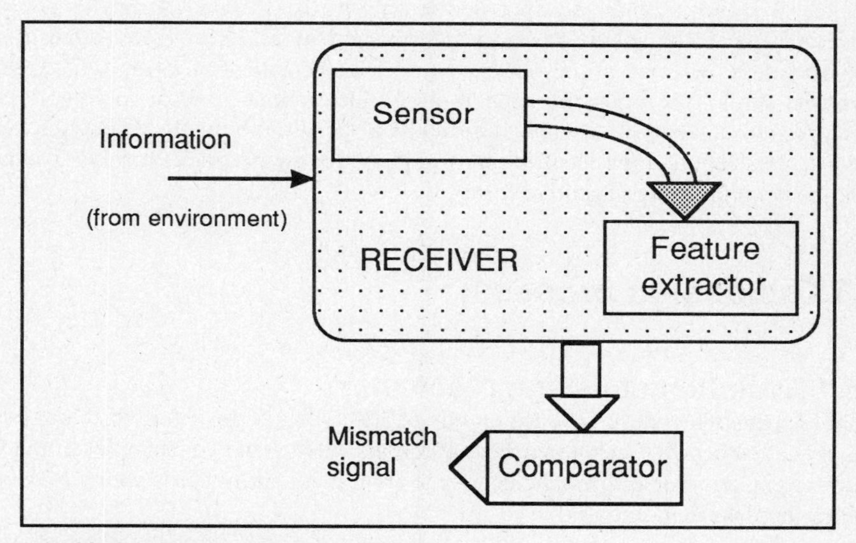

Figure 1.3 Model of problem solver

1. *planning actions* to achieve the desired goal;
2. *trouble-shooting*; correcting deviations in performance from the desired goal.

From the point of view of the professional engineer, planning and trouble-shooting are the two primary forms of problem-solving activity. (The first of these is our concern in engineering design.) However, to support primary problem solving, there are three important types of subsidiary problems which the problem solver may have to deal with in order to achieve the ultimate goal. They are:

1. problems of *resource allocation*, arising from difficulties experienced by the problem solver in matching the resources required to solve the primary problem with the resources available;
2. problems of *information acquisition*, arising from gaps between the information required by the problem solver and the information available;

3. problems of *values*, arising from difficulties experienced by the problem solver in assigning values to the goals to be attained.

The above discussion has been based on the type of activity engaged in by the problem solver. Hinton (1968) and other psychologists describe these activities as processes and classify human problem solving according to the process variables present. An alternative method of describing and classifying problems has been proposed by Ray (1955), Reitman (1964, 1965), and Krick (1969), based on task variables independent of the problem solver. They suggest that in any problem there is an *initial state* of affairs which may be represented by the components of a vector A and a *final state* of affairs represented by a vector B. The *problem* is concerned with the *means of transforming from A to B*, denoted $\twoheadrightarrow$.

Initially there may be some information about B. According to the nature of the problem, B may be completely specified or only partly specified. A may be completely specified, partially specified, or not specified at all. The rules governing the transformation $\twoheadrightarrow$ may be *well-defined* (as in mathematical problems), *ill-defined*, or *unknown*. While this argument helps to clarify ideas about problem solving, its further development runs into philosophical difficulties because it blurs the distinction between a physical system and the idealization of that system in the problem solver's mind (the mind-body controversy in another form).

1.3 The design process

1.3.1 Evolution of design problems

What happens when designers design things? Just what is the nature of this mysterious activity called design? To answer these questions we look at two examples of the way in which engineering design problems emerge from community needs which may be only vaguely apprehended initially.

The problem

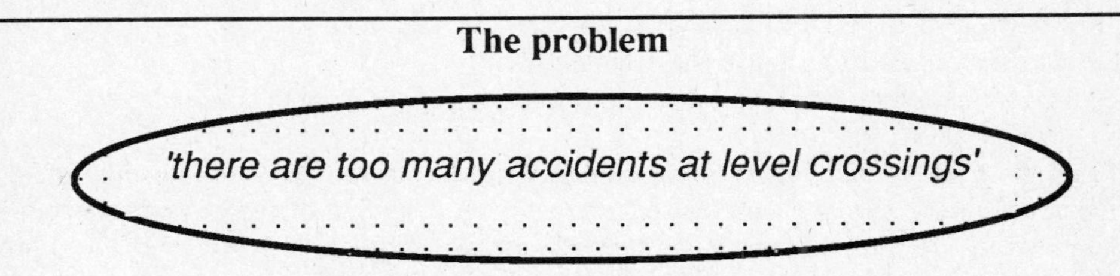

The above *'problem statement'* expresses some vague dissatisfaction with current conditions.

From this *'problem statement'* a *'need'* is established, and this *'need'* is some means of overcoming or reducing the level of dissatisfaction.

Example 1.e.1 *Accidents at a level crossing*

In many parts of Australia the land is flat and there are level crossings where roads and railroad lines cross. The intersection of two characteristically different modes of transport may be a stimulus for engineering problem solving, as the following notes indicate.

Railroad crossing problem -
Some initial questions:

- *How serious is the problem?*
- *How much is it costing us?*
- *How much can we afford to spend fixing it?*
- *Where do we put our most effective design dollar?*

Quantify the problem

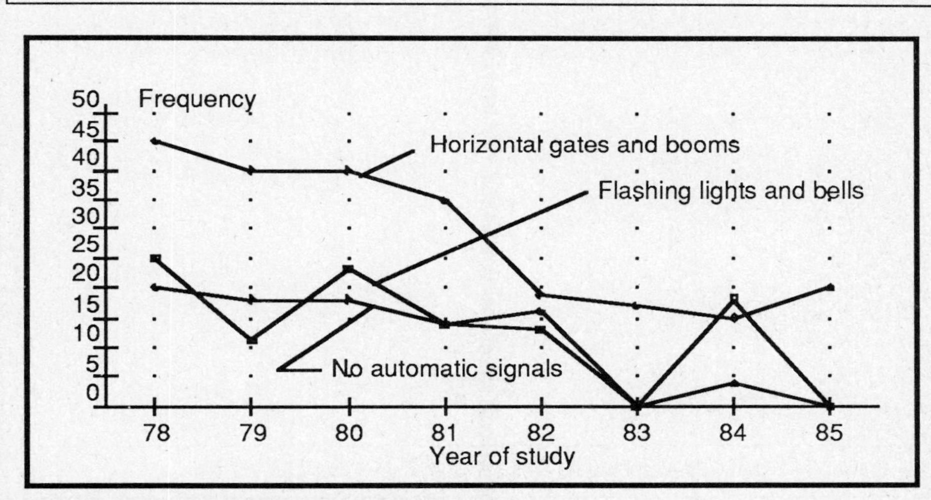

Figure 1.4 Railroad crossing property losses

Note: major concern was expressed in the press about this problem in 1981. The level of concern, public awareness and some fixes may have modified the problem since then.

Data for last four years (1982-85)	*Mortality*	*Property damage*
• *horizontal gates and booms(g&b)*	5	58
• *flashing lights and bells(fl&b)*	8	16
• *no automatic signals(nas)*	0	11
Total	13	85

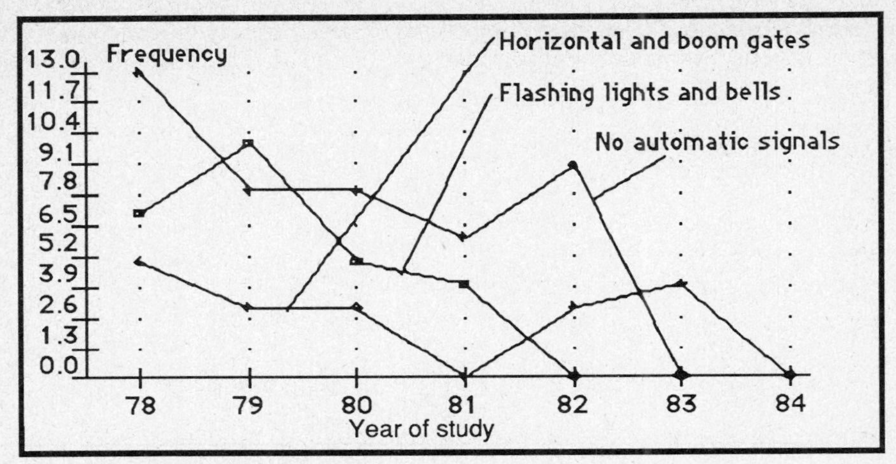

Figure 1.5 Railroad crossing deaths

Number of this type of level crossing		Data/1000 vehicle crossings/annum: Property losses	Deaths
• g&b	181	80	6.9 worst
• fl&b	416	9.6	4.8
• nas	2,581	1	0

Problem elements to be explored

1. The driver:

- possible over familiarity;
- ability to concentrate for long periods; driving on straight bitumen highway, for example;
- response to signals; visual, tactile, such as the road surface for example, auditory;

2. The crossing:

- special geometry - position of the sun in the morning and afternoon;
- presence of other highways;
- levels of illumination.

3. Other:
- institutional barriers to concerted action;
- advertising campaign to arouse public awareness;
- possible separation of the two traffic streams over/under pass.

GOAL

Design an early warning system for level crossings

Objectives:
- suitable for horizontal gate and boom type crossings;
- cheap;
- reliable;
- easy to maintain;
- fail safe.

Other sample problems

The following problems also require '*quantification*' before a '*goal*' may be '*distilled*' from them:

- There is never enough space on faculty notice boards.
- My arthritic aunt can not open milk cartons, soft drink bottles, and plastic food wrappers.
- Underground car parks always smell of gasoline and exhaust fumes.

You can probably generate many other problem statements yourselves. Be warned however! Just because you can put a problem into words does not necessarily guarantee a solution.

Example 1.e.2 *Electric water heater*

Sometimes the engineer is called in when ideas have '*jelled*' to the point where the general nature of the device to be designed has been decided. In this example we suppose that market research has identified a need for a small electric heater and that our designer hero has started to wrestle with the problem. What does our hero do ?

A typical design problem

Design a new type of electric water heater
- must be small;
- must fit into average family car luggage space;
- must be powered by 12 volt battery (direct current).

MATERIALS
- currently available?
- conventional choices?
- exotic choices?

(Small electric water heater)

Small:

- luggage space size?
- how much water do I need at a picnic?
- what are available Thermos sizes?

Electric:

- what types of heating elements are available/in use now?
- is there a standard size?
- can they work with 12 volt d.c.?
- what is the power drain required?

Water heater:

- how will it operate?
- how do we get water in/out?
- skin temperature?
- safety cutouts?

Other points to watch out for:

- one can lose sight of objectives;

- legal implications;

- human and public relations.

Reprinted by permission of the Council of the Institution of Mechanical Engineers from *Lighter Engineering* by R.L. Clarke.

1.3.2 Operational model of the design process

The two examples deal with the inception and evolution of design problems, and provide some insight into the nature of the intellectual activities we call *design*. We now go further and make a list of all the activities in which the designer engages and we express them in the form of an operational model of the design process. This model describes in words the general sequence of operations by which a designer arrives at a final solution. The amount of detail in the model should be sufficient to give an adequate description of the design process without confusing the reader.

1. *Recognition of problem*
The stimulus to engineering design is some human need which has to be recognized. The ultimate goal is the satisfaction of this need; this goal can and should be stated in one sentence without any implication of the means of achieving it.

2. *Definition of problem*
The engineering designer has to develop the following detailed specifications of the task to be accomplished, as far as possible in quantitative terms:

2.1 establishment of objectives and performance criteria — their relative importance;
2.2 identification of the resources available - space, time, money, people, existing knowledge, and physical facilities;
2.3 establishment of the boundaries of the design — the way the problem is to be isolated from its environment;
2.4 identification of the array of subproblems likely to arise during the course of the design'priorities' key issues.

3. *Exploration of problem*

3.1 information search;
3.2 assumptions and estimates to cover gaps in the information available;
3.3 plan of campaign, design strategy - the order in which the subproblems are to be tackled.

4. *Search for alternative proposals*[1]

4.1 search through existing solutions to design problems - technical literature, libraries, patents, manufacturer's catalogues, own experience, other people's experience;
4.2 creation of new proposals.

5. *Evaluation and decision making*

5.1 prediction of the outcomes of implementing each proposal, and the consequences of alternative courses of action;
5.2 analysis of the feasibility of alternative proposals;
5.3 evaluation of the worth of outcomes of feasible proposals by the application of: (a) criteria already established in operation 2 (definition of problem) above; and (b) criteria arising from differences between specific proposals.

6. *Specification of solution*
Progressive synthesis and analysis of alternative proposals leads finally to the optimum solution. Typically, there will be many iterations of synthesis and analysis before the final solution is reached.

---

[1] 'Proposal' denotes a proposed solution of the design problem. It would be ambiguous to describe operation 4 as search for alternative solutions' since some of the so-called 'solutions' may later prove to be unfeasible.

7. *Communication of solution*

The solution to the design problem is recorded and communicated to other people by means of:

(a) sketches and drawings,
(b) written specifications of materials and processes,
(c) scale models, photographs,
(d) computer data links.

Comments

1. Engineering design proceeds from an abstract statement of goals to the final specification of hardware in relatively concrete terms. During this process new information is continually being generated by the designer, who continually re-examines previous findings and decisions in the light of the new information gained and revises them where necessary.

2. Designers (and engineers in general) have to become accustomed to making probabilistic decisions. There are always uncertainties in the data available, doubts about the validity of assumptions on which design calculations are based, or doubts about the validity of laboratory tests for predicting the performance of the final design.

3. Acceptance of the above operational model as a theoretical framework for formal discussion of engineering design should not prevent us from surmising that the problem-solving methods adopted by designers in particular cases may be of an informal or intuitive nature. This is most likely to be the case in simple applications where the amount of information to be processed is small.

4. Nor should we lose sight of the dynamics of problem-solving behaviour — the possible feedback and feed-forward of information between different stages of the design process. For example, the designer may change the objectives if he/she finds there are insuperable difficulties to achieving them. As a result of their study of an innovative design problem, Frichsmuth and Allen (1969) concluded that an important way in which engineers could respond to difficult design problems was by modifying objectives and setting lower standards of performance.

1.3.3 Flow chart

Another model of the design process is provided by the flow chart (see Figure 1.6). The chart has been constructed to show the important steps in engineering design without being overloaded with detail.

As the final question mark in Figure 1.6 indicates, the design process is not complete until the solution has been implemented. Its effectiveness can then be monitored and the information so obtained used to initiate another design cycle, as illustrated in Figure 1.7.

1.4 The design hierarchy

It is often helpful to consider a complex design as an engineering system, and then to develop strategies for dealing with the elements of the system in an orderly way. A system is a set of objects (elements) with known relationships between the elements and between their attributes. *Examples*: Power station, water supply scheme, city, motor car.

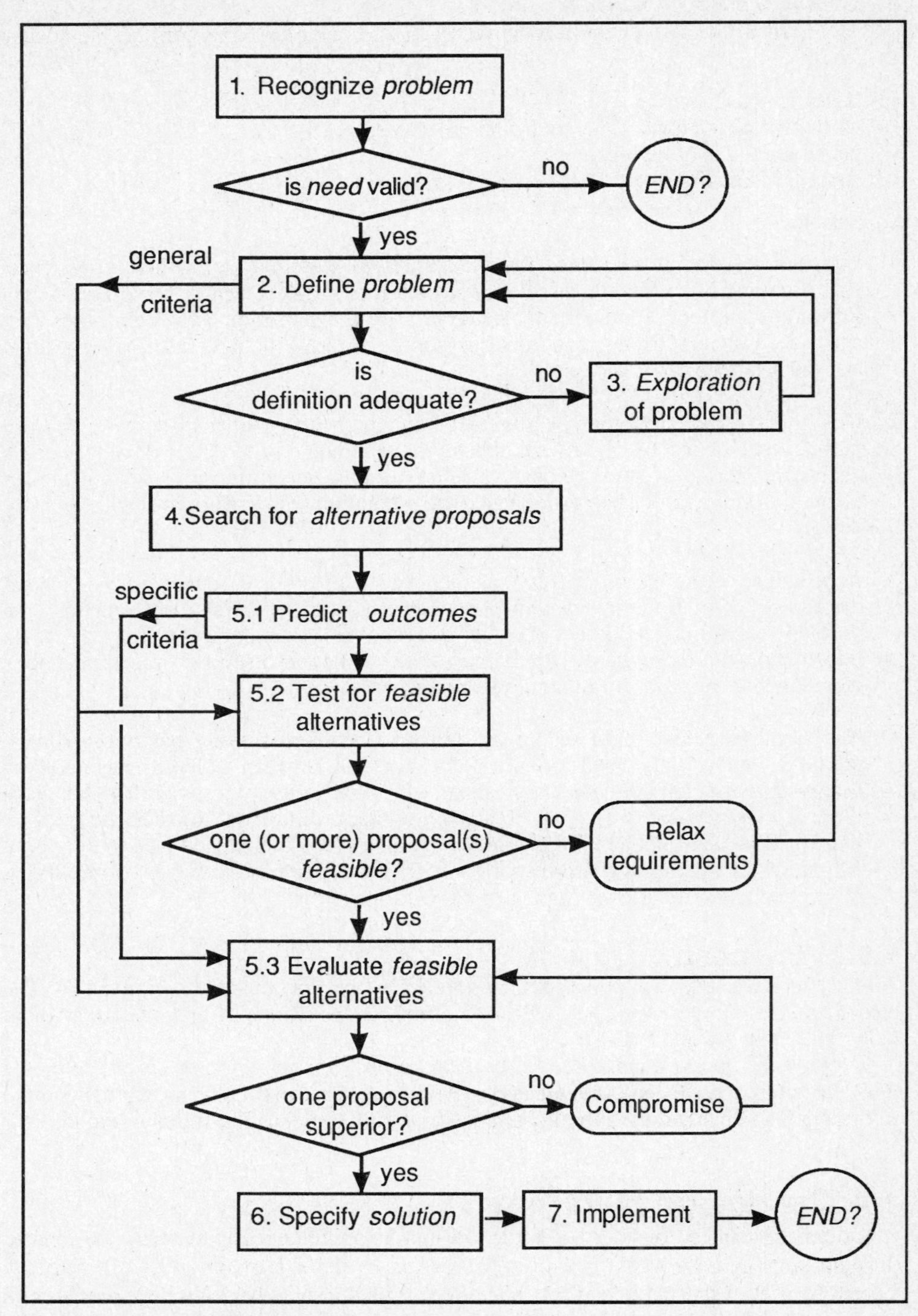

Figure 1.6 Flow chart of the design process

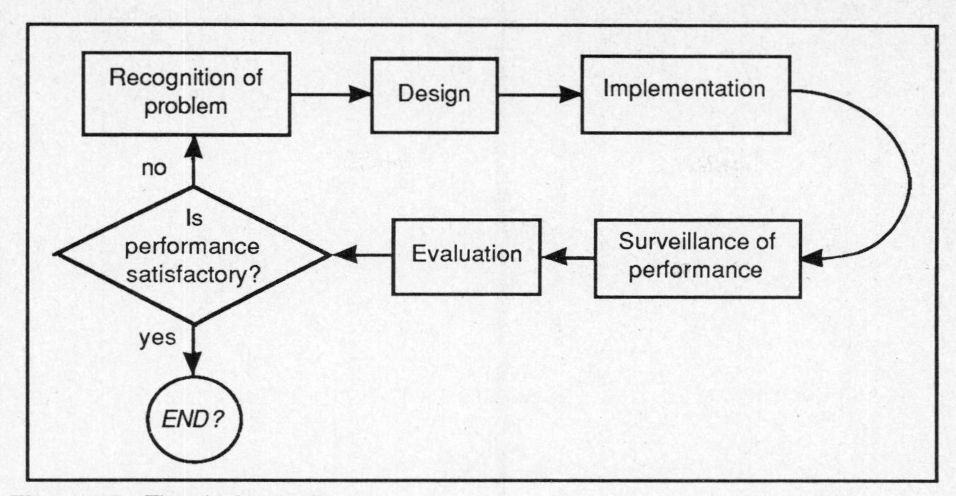

Figure 1.7 The *design cycle*

The environment of a system is the set of all objects external to the system, whose behaviour affects the system and which are in turn affected by the behaviour of the system. There will be exchanges of matter, energy, and information between the elements of the system and its environment.

Engineering systems commonly operate in a state of imbalance with self-correcting features being an integral part of the design. The block diagram of a simple system with feedback is shown in Figure 1.8.

System is a multilevel concept. The number of levels in the design hierarchy is a measure of the complexity of the design. Thus, as shown in Figure 1.9, there are four levels in the design of a motor car, but only two for an electric motor. Complex engineering designs with more than five hierarchical levels are rare.

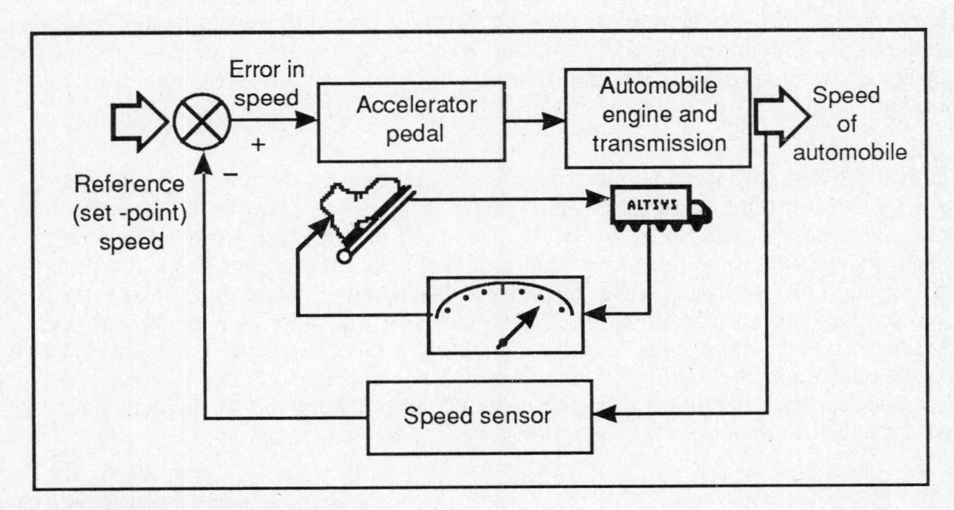

Figure 1.8 System with *feedback*

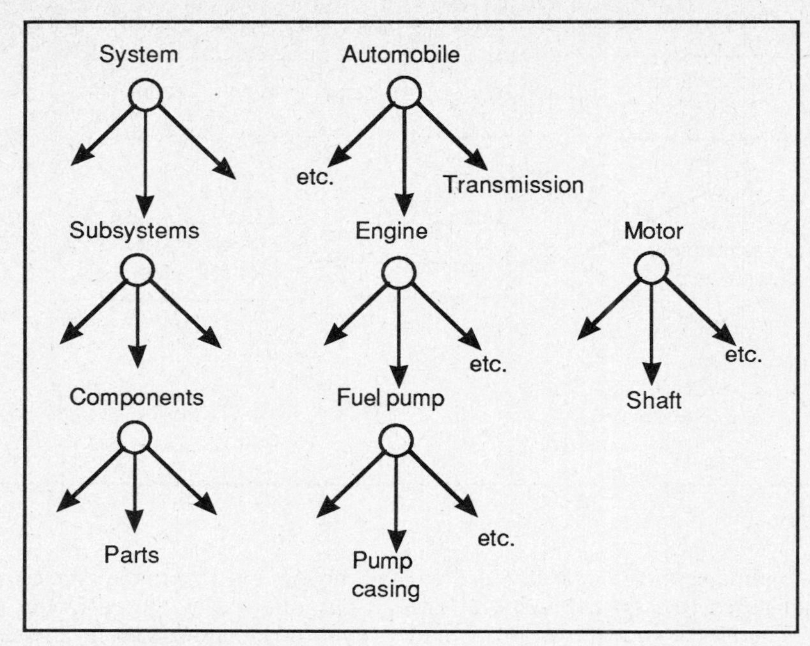

Figure 1.9 The *design hierarchy*

1.5 Review of design thinking

1.5.1 Design thinking and everyday thinking

Before we conclude this introductory discussion of engineering design, there are some important matters to be reviewed concerning the nature of design thinking and the strategies used to solve quantitative design problems.

Successful engineering design depends on modes of thinking distinctly different from those we use in everyday life. The existence of these differences makes design an intellectually demanding activity. To see that this is so, let's look at some of the characteristic features of *'design thinking'* which distinguish it from *'everyday thinking'*.

1. *Goals*

Engineering design is goal-directed; the designer is consciously aiming to satisfy some need or want for other engineers or for society at large. For example, society says *'We want more energy'*, and the engineer replies *'I will design these energy converters to satisfy your needs'*. The purposive nature of the work affects everything the designer does. By contrast, everyday life is much more disorderly; a lot of things just happen to us or are beyond our direct control. Despite our best intentions, few of us consciously direct our lives towards some long-term goal. On a more mundane level, we don't go around each morning saying *'I have a problem, how am I going to travel to the University today*?' We don't waste time in making problems out of routine events; we just catch the bus or get out the car, or whatever.

2. *Complexity*

Much of everyday thinking is personal to the individual. We can narrow our thinking to suit ourselves and concentrate on small, perhaps minor, aspects of our life and work. As long as we keep our thoughts to ourselves no one will tell us where they think we are wrong. But design problems are for real. They affect many people. Designers cannot act in isolation, they must take into consideration many parallel strands of thought and argument, and knit them together into one complete solution.

The contrast is shown in Figure 1.10. Much of everyday thinking is linear, based on the individual frame of reference of the person concerned, and therefore partial. On the other hand, designers are synthesizers. Ideally, they identify all relevant factors and they apply all relevant criteria when making decisions. They cannot afford the partial, blinkered thinking of everyday affairs.

3. *Time scale*

Life is a *'one-off'* experience we are not free to experiment with it as we like. Everyday thinking is primarily concerned with the here and now. On the other hand, designers plan for the future; they are concerned with future embodiments of present ideas. They are completely free to carry-out *'thought experiments'* — to play around in their minds with ideas and fancies which may not even be capable of practical expression. Often it is necessary for them to do this in order to hit on the best way of achieving their goal.

Designers have to predict the outcomes of various courses of action open to them. If they find that some outcomes are hazardous, unsafe, or uneconomical, they can back-track in their thinking and start again from new premises. Trial and error is a characteristic mode of progression, as indicated in Figure 1.11.

The design process is an aggregation of a very large number of these feedback loops, as the designer gradually converges on to the final, detailed solution.

4. *Precision*

The engineering designer has to communicate with other engineers, either to persuade them to accept plans or to give instructions on implementation. This demands a high degree of precision and a readiness to think in quantitative terms — to put numbers on things.

To round off this discussion, let's not forget that *design thinking* also resembles *everyday thinking* in some important respects.

Firstly, design decisions have often to be made on the basis of inadequate or incomplete information. Life doesn't serve information up to us on a platter, so we often have to give something a go! Frequently it is better to make a decision, any decision, rather than sit around twiddling our thumbs, Hamlet-like, waiting for more information

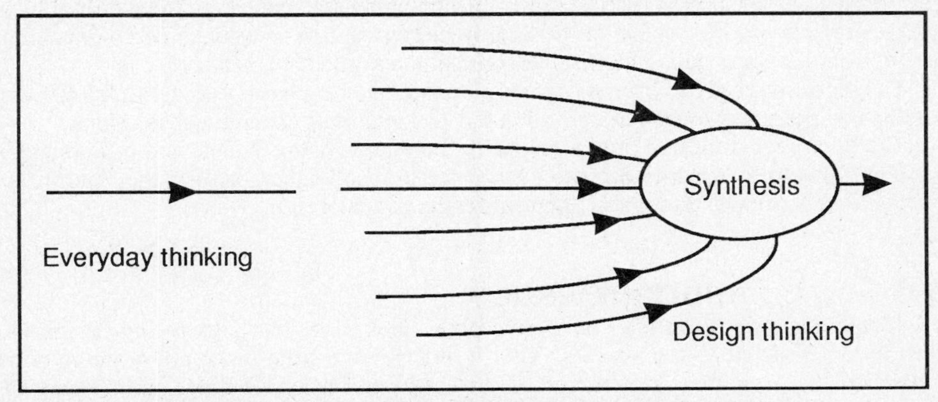

Figure 1.10 The Contrast between *everyday thinking* and *design thinking*

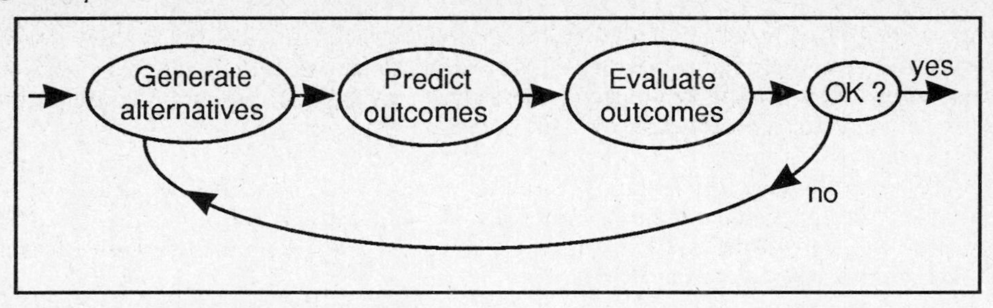

Figure 1.11 Feedback loops in design

to come to hand. An example from politics: The Federal Treasurer has to make many decisions in the budget with only a limited knowledge of how they will affect the Australian economy.

Secondly, we generally try to keep our options open for as long as possible, to explore alternatives as far as we can before committing resources to one specific course of action. Edward de Bono (1976) has drawn attention to the use of 'porridge words' — vague and general words which we use to help keep our thinking going while we search around for more specific ideas or explanations. For example, designers are often faced with many uncertainties; they have to translate porridge words and phrases like '*hairy*', '*a rough chance*', and '*a good bet*' into precise recipes for action. They may do this in an informal way by exercising their own subjective judgment, or they may do it formally, as will be explained in Section 7.3.

1.5.2 The dilemmas of design

Engineering design begins with the identification of some need or want in relatively abstract terms and it ends with the statement of a detailed plan for the manufacture or construction of some device or system which satisfies that need. The process of translating '*abstract*' need into '*concrete*' hardware inevitably confronts designers with some major dilemmas.

On one hand they have to be capable of the broad view, the vision that takes account of second- and third-order effects, and on the other hand they have to be precise and detailed in their conclusions.

They have to be imaginative, flexible and open minded in generating and handling new ideas, but they also have to be critical in analyzing and evaluating the worth of these ideas. They have to be capable of thinking in these two different modes and of switching at will from the imaginative mode to the critical mode and back again.

They must be ready to try new procedures and methods, but not afraid to fall back on established precedents when these lead to the best and most economical solutions.

The dilemmas of design are inescapable. Engineers, both students and professionals, have to draw on their education and experience to resolve them as best they can. It is a process that continues throughout their professional careers.

1.6 Some important issues

The preceding discussion may have given the impression that engineering design is a straightforward, clear-cut activity. This is not so. We now draw attention to some important issues which are sources of difficulty and often of great controversy (de Neufville and Stafford, 1971):

1. Can a well-defined set of objectives be established in the first place? Is there a consensus among the different people or groups of people concerned? Different points of view regarding a new freeway may be held by: (a) its planners, (b) the owners of the land it will traverse, (c) the people who will live near it, (d) the drivers of vehicles who will use it.

2. What measures of effectiveness (criteria of benefit) are pertinent to the objectives? What scales should be used for these measures? How does one allow for intangibles such as aesthetics and the subjectivity of people's responses to them? More generally, is it possible to quantify human behaviour and is it legitimate to do so?

3. When alternative proposals are being generated, should step 4.1 precede step 4.2 in the design process? Or would this give rise to preconceived ideas in the mind of the designer and inhibit the creation of fresh or original approaches? What is the most efficient technique of searching for candidate solutions?

4. How should the future be predicted? Does the technique of simulation used by the designer rely on many simplifying assumptions? Do all concerned agree with these assumptions and extrapolations into the future?

5. If compromises are to be made, how does one arbitrate between the claims of conflicting requirements, many of which may be multidimensional?

6. Can all the consequences of implementing a design be foreseen? Can second- and third-order effects be allowed for by designing a sufficiently robust yet flexible system?

These questions arise time and again in the practice of engineering design. Some answers are given in the chapters which follow, while other answers will be found by applying the general methods described in this book to particular cases of interest.

Chapter 2

The divergent phase of the design process

Imagination is more important than knowledge.
Albert Einstein

Concepts introduced	goal; objective; criterion; constraint; creativity.
Methods presented	input-output analysis; initial appreciation; aids to creative effort.
Applications	(a) photocopier for academic use; (b) design of a wheelchair.

2.1 Recognition of design problems

Psychological research has shown that people vary in their sensitivity to the problems inherent in a situation but which are not explicitly stated (Guilford, 1967). This has important implications for engineering because in practice engineering designers frequently receive very vague statements of problems from their employers or clients. Their first responsibility therefore is to identify the problems which are not explicitly stated in the briefs given to them but which are implicit in the briefs.

There may also be social interactions between designers and the people who refer problems to them. Such interactions most frequently arise when the client is a layperson with little knowledge of engineering. Mitroff (1968) describes an example of an engineering consultant to a nuclear physics laboratory who (a) had to defend himself against know-all scientists who over-defined the technical problems they asked him to solve, and (b) had to cope with inexperienced physicists who didn't know what they wanted their experimental equipment to do anyway.

The work of some engineering designers impinges on many members of the community, for example, the designers of major structures such as bridges, communication systems and consumer products. The opinions and attitudes of many people may affect their work. They may have to depend on market surveys and interviews to determine whether or not other people's views of the problem are the same as theirs. Sometimes existing social institutions place a barrier between the designer and all the people who ultimately will be affected by or will respond to the designer's efforts. It may then be a matter of personal ethics whether an individual designer accepts this barrier or tries to circumvent it.

2.2 Definition of design problems

2.2.1 General

When a specification of the task to be performed is drawn up, it is essential that the goal be stated in a general way without reference to hardware. The initial statement should preferably consist of one sentence.

Far too often the inexperienced designer rushes into a new problem and concentrates on one possible solution, with no appreciation of the wide range of possibilities available. If the design of a lawnmower is being considered, for example, it should be realized that the goal of the endeavour is to keep lawns neat and tidy, and the lawnmower and its designer can be replaced by the scientist who designs a slow-growing grass. The number of possible solutions increases with the generality and breadth with which the problem is initially stated; it decreases with the number of restrictive and inhibiting words used.

After the goal of the endeavour has been clearly established, more detailed specifications can be drawn up. However, it must be realized that engineering design is a dynamic process. As a design progresses, the feedback of new information enables the designer to continually refine objectives; in fact the value system and the physical system are under simultaneous development. Gradually it becomes clear where there are difficulties in satisfying conflicting objectives, and the nature of the compromises to be made emerges. At all costs the designer must resist the temptation to try to solve the problem before its true nature has been grasped. The adoption of premature *'solutions'* invariably results in patch-up jobs and much wasted effort.

2.2.2 Checklists of objectives

1. *Product design*

In developing the specification of the task to be accomplished, the product designer must consider all the objectives which have to be satisfied, and must rate the relative importance of these objectives, that is, whether they are essential, highly desirable, desirable, or marginal.

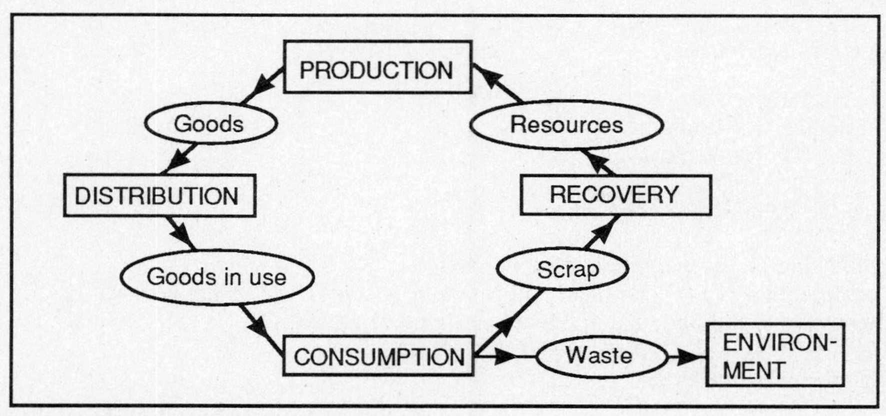

Figure 2.1 The production-consumption cycle

Objectives can be classified under three headings:

1. performance (effectiveness);
2. appearance (aesthetics);
3. cost (efficiency in use of resources).

In engineering design, it is nearly always the case that the cost of the product being designed is to be a minimum consistent with performance and appearance at an acceptable level of quality. A checklist of objectives is obtained from consideration of the production-consumption cycle (Figure 2.1). Industry produces and commerce distributes the goods and services that people consume. After consumption, the waste products are removed — usually they are destroyed and sometimes they are salvaged.

Design for consumption:

1. ease of handling and installation;
2. ease of operation; suitability of the operating characteristics of the product — quietness, stability, sensitivity, compatibility with the environment;
3. ease of maintenance;
4. durability, long service life;
5. reliability, low maintenance costs and short down-time;
6. efficiency, low operating costs;
7. weight;
8. volume, plan area, frontal area;
9. safety;
10. flexibility.

Design for production:

11 ease of production;
12. use of available resources — manufacturing plant, raw materials;
13. use of standard parts and methods;
14. use of facilities for quality assurance;
15. reduction of rejects and scrap parts and material.

Design for distribution:

16. ease of transport;
17. suitability for storage;
18. suitability for display.

Design for retirement of product:

19. matching of physical life and service life;
20. replaceability of (a) complete product, and (b) short-lived components;
21. recovery of reusable material and long-lived components.

Design for environment:

22. effect of product on the natural environment during its service life;
23. effect of product on the natural environment after it has been taken out of service.

2. *System design*

Many of the objectives in the above list are also relevant to the design of engineering systems (e.g. in the fields of transport, water supply, waste disposal, communications). The construction of large engineering systems consumes a high proportion of a community's resources, and the benefits and costs have therefore to be carefully examined. Moreover, such systems are likely to have a significant impact on the physical environment. Environmental and ecological objectives have to be added to our check-list even though they may be difficult to quantify. The types of question that might be asked during the planning of a new dam are:

- How many species of plants and animals will be destroyed by the water being stored?
- How many of these species (if any) are unique to the region concerned?
- Will any interesting, even unique, geological features be submerged?

2.2.3 Design objectives, criteria, and constraints

All successful designs possess a distinctive unity and coherence which defies analysis. Nevertheless, for the purpose of this discussion it is necessary to separate out some of the key issues and examine them in sequence.

Designers continually have to ask themselves where they are going; they have continually to *monitor* their *objectives* and *progress* towards those objectives. The evaluation and comparison of alternative proposals implies the existence of standards of reference, yardsticks, *scales for measuring advantages and disadvantages*. These scales are the *design criteria*.

While *objectives* are formulated in purely verbal terms, for example *'the ship's propeller must resist corrosion by seawater'*, the corresponding *criterion* (or *criteria*) provides a measure for comparing the performance of different materials which may be used in the design. For example, a propeller made of alloy A has an expected service life of X years or its material corrodes at a rate of Ymm per year in seawater. The objective of *'corrosion resistance'* gives rise to the criteria *'expected service life'* and *'rate of loss of metal by corrosion'*.

Another example may be helpful. Consider the bushwalker who plans to walk to some distant point X. This then is the objective. There may be several tracks available hence alternative proposals need to be considered. One path may be longer than another, one may require more muscular effort going up and down hills, yet another may traverse a particularly beautiful stretch of country. The criteria which the bushwalker uses to arrive at a decision on which path to take are: (a) time to reach X, (b) muscular effort expended, and (c) scenic beauty. These criteria are measured on different sorts of scale; *time* provides an objective, continuous scale; *muscular effort* can be measured objectively also, but the walker would probably rely on a subjective assessment of the scenic beauty; however, 'a thing of beauty is a joy forever' (Figure 2.2 illustrates these ideas).

Engineering design is a dynamic process! Designers continually review the aims and scope of their endeavours. Young engineers early in their careers tend to be asked to solve fairly specific problems whose boundaries are well-defined. Senior engineers, however, have to accustom themselves to working in vague and ill-defined situations where they have to devote considerable effort to defining the problem boundaries. No matter what the scale of the design, the dictum of the American architect Eliel Saarinen is worth bearing in mind.

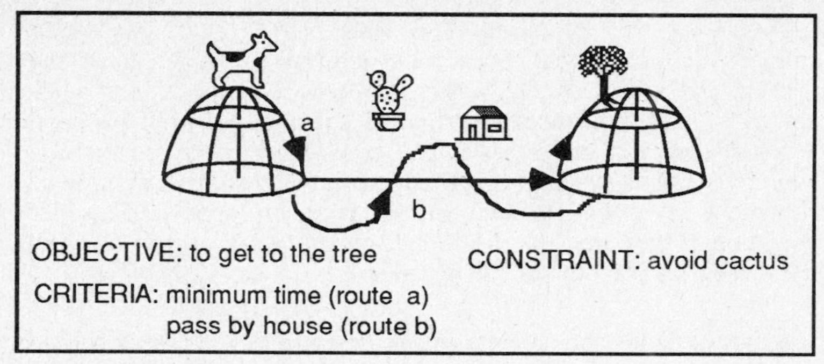

Figure 2.2 The dog and the tree

'Always design a thing by considering it in its next larger context —
a chair in a room, a room in a house,
a house in an environment, an environment in a city plan.'

2.2.4 Input-output analysis

Both Asimow (1962) and Krick (1968) recommend input-output analysis as a technique for helping to define design problems. It provides a framework for guiding our thinking about the design of engineering systems and products. The general approach is shown in Figure 2.3 and an example of its application to product design is given in Example 2e1. The product is regarded as a device for transforming certain inputs into outputs, some of which are desired and some of which are not.

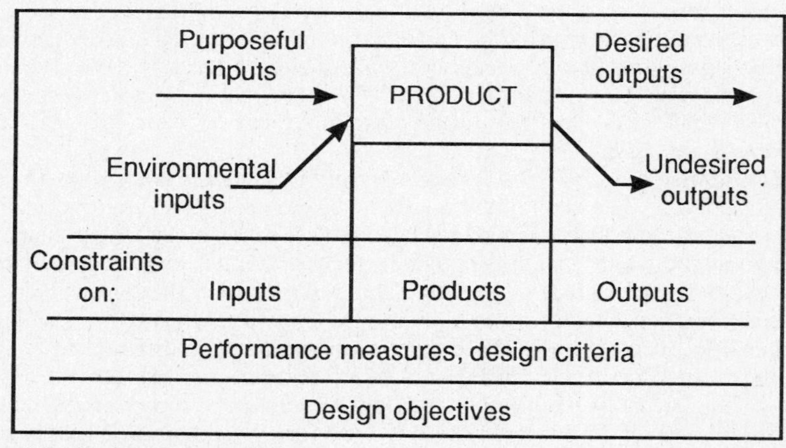

Figure 2.3 Input-output analysis

Example 2.e.1 *Photocopier*

To illustrate input-output analysis, we consider the problem of designing a photocopier for use by an engineering research and development organization where multiple copies are required of: (a) research reports, (b) engineering drawings and diagrams, (c) papers for meetings, minutes, copies of letters and correspondence for filing and future reference. The copier is to be placed in a room used for research meetings and '*think tank*' sessions.

Overall design objectives: Rapid production of high quality copies of reports, papers, correspondence, drawings and diagrams at reasonable cost, with extended periods of trouble-free operation, machine to be easy to use and to be of neat and pleasing appearance.

Desired output: Copies of documents inputted to the machine.

Undesired outputs: Smudged or spoiled copies; frustrated users delayed by machine malfunction; heat from the machine.

Purposeful inputs: Documents to be copied; sheets of copying paper; electrical power; person to operate the machine; funds to cover cost of installation and operation of the machine.

Environmental inputs: Chalk dust from blackboards used in '*think tank*' sessions.

Constraints on outputs: Copies produced should be of acceptable quality (a subjective judgment); machine to be capable of outputting copies on A4, US standard letter and foolscap size paper; at least one copy per second to be produced during multiple copying of a document.

Constraints on inputs: Size of documents to be copied would ordinarily be A4, US standard letter, foolscap or A3; machine to be capable of handling intermediate sizes up to a maximum of A3; costs of installation and operation of the machine to be within the research organization's budgets; a person of average intelligence to be able to learn to use the machine with less than five minutes instruction.

Constraints on product: Machine capacity to be at least 2,000 sheets of copy paper; not more than one malfunction or misfeed per 1,000 copies produced; size of machine to be less than allocated floor space available.

Performance measures and design criteria:

(a) objective performance measures: number of photocopies produced per second; number of malfunctions per 1,000 copies produced; learning time for intelligent user; capital cost of installation; operating cost per 1,000 copies.

(b) subjective assessments: quality of copies; pleasing appearance of machine.

Comments

Input-output analyses are chiefly of use in applications where sufficient experience has been gained to enable the different inputs and outputs to be identified.

Krick's example (Krick, 1968, p. 118) is open to a number of criticisms. Firstly, he includes a section on '*solution variables*' and thereby mixes up the solution phase of the design process with the earlier phase of problem definition. In everyday life we can mix up the different phases of problem solving without coming to much harm. In engineering design, however, it is essential that a problem be properly defined to start with, if only to prevent time being wasted solving the wrong problem. Secondly, the meaning of '*input*' and '*input variable*' is not clear. Furthermore, are A and B states of a physical system or states of mind of the designer? Lastly, he does not show how the criteria (measures of effectiveness) are quantified, that is, on what scales they are to be measured. Despite these reservations, Krick does show how the engineer may come to grips with a vaguely defined problem.

2.2.5 Matching

Thought should be given at an early stage of a design to a type of problem which frequently arises subsequently, namely '*matching*'. Jones (1966) distinguishes between '*flow*' and '*associative*' systems in engineering. In the former, there is a flow of, for example, mass, energy, or information from one element of the system to another (e.g. power stations, transport systems, communication systems). In the latter, elements are associated together in some definite spatial relationship but there is negligible flow of anything from one element to another. An *automobile* may be regarded as an *associative* system consisting of engine, transmission, chassis, passenger compartment and suspension.

Matching is the problem of ensuring that different elements of a system are compatible with each other and that the system and its elements are compatible with the environment. In associative systems the dimensions of adjacent elements must be compatible; while in flow systems the operating characteristics of connecting elements must be compatible. For example, in a gas turbine engine the operating characteristics of compressor, combustion chamber and turbine must all be compatible to prevent compressor surges and flame-outs. In both associative and flow systems the designer has to take great care in assigning ranges of tolerances to the design variables under her/his control.

2.2.6 The initial appreciation

Observations by the authors and others of the working methods of professional engineers have led to the recommendation that an '*initial appreciation*' be made at the start of a new design of any complexity. The appreciation should cover most of operations 2 and 3 in the design process (see Section 1.3). The aim is to define the starting point of the design activity and to plan the work required to bring it to a successful conclusion.

An initial appreciation covers the following key matters:

1. objectives to be satisfied and their relative importance — whether essential, highly desirable, desirable, or marginal;
2. criteria to be used in evaluating alternative designs — measures of effectiveness, measures of the resources used, and benefits and costs;
3. information required as a basis for the design and the sources from which it may be obtained;
4. problems foreseen to arise in the course of the design, especially those arising from conflicting objectives;
5. plan of campaign or strategy for tackling the design, set out according to a scheduled timetable or '*critical path*' network.

Example 2.e.2 *Self-propelled wheelchair*

To illustrate the application of the foregoing ideas and how engineers plan their approach to a design problem, we will look at a particular example, where some decisions have already been taken about the nature of the solution. Suppose then it has been decided to design a self-propelled wheelchair for people who are unable to walk because of their injury or disease, but who retain full use of their mental faculties and their upper bodies, including their arms.

OBJECTIVES	CRITERIA
1. EASE OF OPERATION BY INVALID	
• ease of propulsion	• forces/torques exerted
• manoeuvrability	• turning circle
• ease of control – steering, accelerating braking, stability	• forces/torques exerted – response time
• chair to ascend/descend ramps, steps, gutters	• slope of ramp; size of step, gutter
• ease of entering/leaving chair	• forces exerted; time taken
• comfort	• subjective rating number/size of bedsores
• range of operation	• distance travelled without attention required to power source
2. RELIABILITY AND MAINTENANCE	
• operation	• forecast number of breakdowns
• cleaning and repair	• mean time to clean and repair
3. DURABLE CONSTRUCTION	
• long life, no parts easily broken due to manoeuvring	• forecast life of chair and components
4. WEIGHT	
• chair to be light to assist portability and propulsion	• weight
5. SIZE	
• passage of chair through doorways • transport of chair in automobiles and public transport	• width, overall dimension, weight
6. SAFETY	
• emergency braking	• braking distance, slope
• protection of invalid in crash overturn	• forecast injuries
7. PROPER USE OF RESOURCES	
• ease of production	• number and complexity of components

• low cost	• manufacturing cost
8. AESTHETICS	
• appearance	• subjective rating

PRIORITIES: Difficult to set priorities until criteria are expressed quantitatively. Objectives 1 to 6 seem to be the most important ones.

INFORMATION REQUIRED

- number of invalids (population)
- anthropometry (shape and size)
- mental/physical capability/disability and the effect of these on the control/operation of chair
- physical environment in which the chair is to be used
  - indoors/outdoors, home, hospital, street, transport vehicle
- geometry of the environment and surface properties
- range of actions to be performed
  - are urinating, excreting, and eating included?
- existing power sources and construction materials

SUBPROBLEMS FORESEEN

- difficulty of matching chair to existing environment; steps, doorways, toilets, vehicles
- diversity of people to be catered for
- difficulty of keeping weight of chair within acceptable limits

DESIGN STRATEGY

1. collect information
2. design chair around invalid
3. check design
   - Does it match the physical environment?
   - Is it light/portable?
   - Is it likely to be cheap to produce?
   - Does it satisfy the other criteria?
4. if the answer to any of these questions is *no*, iterate until a compromise solution is found

2.3 The search for alternative solutions

2.3.1 The engineering repertoire

There is a vast repertoire of existing solutions to design problems contained in the technical literature, patents, manufacturers' catalogues, standards, and the designer's personal experience — often sufficient to generate several alternative proposals for investigation and evaluation. To help the designer access this repertoire, researchers have classified the functions performed by engineering devices and artefacts and the physical means used to achieve these functions, thus providing the software and hardware of engineering design. For example, common functions would include transmitting power, storing energy, and transporting materials. Chapter 5 of Pahl and Beitz (1984) gives advice on these and related matters, while a quantitative overview of the characteristics of fluid flow machinery is provided by French (1985, p. 111).

In novel problems with few or no precedents, the designer may fail to find an answer in the repertoire of existing practice. Deliberate steps then have to be taken to create new proposals; intellectual muscles have to be flexed, and creative thinking skills must be exercised.

2.3.2 Creation of new proposals

The word '*creativity*' is often used as if it were some inherent quality which people possess in varying degrees. Here we are not interested in abstract definitions but in the demonstrable ability of a designer to generate new and useful ideas for accomplishing engineering goals, that is, creative effort. Presumably heredity, environment and prior experience impose some limitations on a person's inventive ability. However, it is probable that few people make full use of their inborn imaginative potential.

The development of our powers of thought and language occurs naturally as we grow up, so much so that we largely take them for granted. Nevertheless, thinking is a skill capable of nurture and improvement. This book is about the skill of designing and its contents must often, as now, overlap with general considerations of human intellectual endeavour. To illustrate some general matters of significance, consider the following problems:

- Figure 2.4(a) shows matches arranged to form five squares. Can you rearrange the matches to form four non-overlapping squares by moving two matches?
- Given six matches forming a regular hexagon as in Figure 2.4(b), can you rearrange them to form exactly four equilateral triangles?
- Given four lengths of chain of three links each, you are to join them into a single, continuous loop. It costs 3 cents to open a link and 2 cents to close one. Can you perform the task for 15 cents?
- Two friends are walking together. They start level with each other and move off on the left foot. One takes strides of 90 cm, the other 60 cm. When will they step off on the right foot simultaneously?

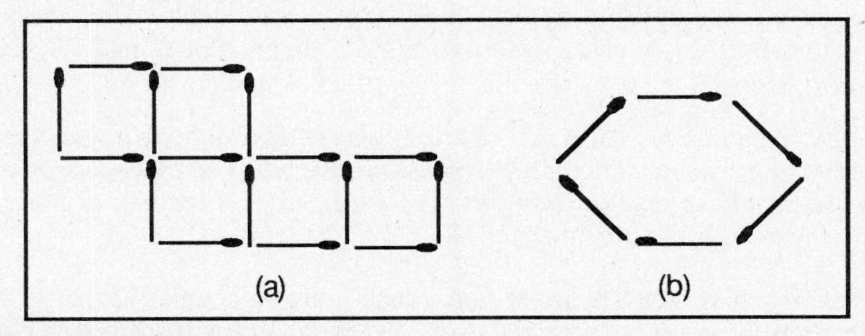

Figure 2.4 Problems with matches

The four problems are well-defined in that the initial conditions given and the final conditions to be satisfied are clear and unambiguous. Yet intelligent people have been observed to vary widely in their responses, both in the methods used to arrive at their answers and the time taken. The differences in people's responses are even greater when the problems presented to them have an open-ended character, where the nature of the final results is not precisely defined. Think about the following exercise and take five minutes to write down your answers before reading on.

'Uses for' exercise

Suppose the 'Darker than Amber' brick company have hired your services as a consultant. They want to expand business and your task is to suggest new ways of using standard house bricks.

Well, how many ideas did you think of? If your ideas included many such as: build a wall, build a hardware store, build a garage, build a shoe store, etc., then you were *fluent* in your thinking, but certainly not *flexible*. A flexible thinker would have suggested a number of different types or categories of brick use, for example, bed warming, water storage, book shelf support, colour gauge, and the new sport of 'brick putting'. Try the brick problem again but listing the attributes or properties to be used in the brick promotion program, for example, weight, colour, rectangularity, porosity, strength, surface texture, hardness, thermal capacity, and thermal conductivity.

To some people the process of generating ideas for solving design problems is fraught with difficulty, while to others it comes as easily as breathing. Why is this? Perhaps the reason is that, as with physical activities we engage in, such as tennis or golf, thinking is a type of skill capable of development. Just as sports champions play with the best sporting equipment, so you, the readers, need the best intellectual equipment to succeed in the game of thinking, particularly when confronted with difficult problems which stretch imagination and creative ability to the limit.

Examples of successful creative effort — some well known, some not — are now presented and discussed. We want to see what we can learn from them about ways of lifting our game.

James Watt. As a young man Watt became interested in improving the Newcomen steam engine when he discovered, while repairing a model at the University of Glasgow, that its mode of operation was extremely inefficient. Power for each stroke was developed by first filling the cylinder with steam and then cooling it with a jet of water; this cooling action condensed the steam, so forming a vacuum behind the piston which was then forced to move by the pressure of the atmosphere. With every stroke of the piston the cylinder was alternately heated and cooled, and calculation showed Watt that this process was very wasteful of the heat supplied to the engine. He reasoned that if he could prevent this loss of heat, he would be able to reduce the engine's fuel consumption by 50 percent, an accomplishment that was obviously worthwhile. After working on this problem fruitlessly for two years, Watt went for a walk one fine Sunday afternoon when in his own words,

> *'The idea came into my mind that as steam was an elastic body it would rush into a vacuum, and if a communication were made between the cylinder and an exhausting vessel the steam would rush into this vessel and might there be condensed without cooling the cylinder,' (in* Smiles, 1904*).*

Frank Whittle. While on a Royal Air Force cadet training course Whittle wrote a thesis on *'Future Developments in Aircraft Design'* (Whittle, 1953). It seemed to him unlikely that the conventional combination of piston engine and propeller would meet the power plant requirements of the high-speed, high-altitude aircraft he had in mind. Thus in the thesis he discussed the possibilities of rocket propulsion and of gas turbines driving propellers. For 18 months after completing the course, Whittle continued to consider the power plant problem in his spare time from his flying duties. He gave much thought to a jet propulsion arrangement in which the propelling jet was generated by a low pressure fan driven by a piston engine, but eventually came to the conclusion that this offered no real advantage over the piston engine-propeller combination. Then suddenly one day the idea came to him of substituting a gas turbine for a piston engine in this arrangement; he had conceived the jet engine.

We note that Whittle's experience is in accordance with a widely held definition of an engineering creation or invention being *'a new and useful combination of existing elements'*. He created a new and useful aircraft power plant by the combination of two existing elements — the gas turbine and propulsion by the reaction of fluid jets. On the other hand, Watt's invention of the condenser can be viewed as a dissociative process in that he separated the function of condensing the steam from that of transmitting motion to the piston.

George Cayley. Cayley's attack on the problem of manned flight in the first half of the nineteenth century gives another example of dissociative thinking. Early in his investigations he took the decisive step in the history of aircraft design of separating the function of providing propulsive thrust from the function of providing lift. In 1800 he wrote:

> *'The whole problem (of manned flight) is confined within these limits, viz. — to make a surface support a given weight by the application of power to the resistance of air. There can be no doubt that the inclined plane, with a horizontal propelling apparatus, is the true principle of aerial navigation by mechanical means'* (*in* Gibbs - Smith, 1962).

Minor Inventors. Allied to the human ability to break existing linkages between ideas and concepts to form new ones is the ability to think metaphorically — to make use of analogies based on the recognition of similarities between apparently dissimilar situations. The inventor of a wine cask tap (i.e. a valve mechanism for discharging wine from a plastic container) has written:

> *'By chance I observed a milk feeder for babies that included a flexible nipple within a cylindrical container. On deforming the nipple I noted the creation of a deformation of the diaphragm adjacent the cylindrical wall, and appreciated that if I inverted the nipple I would have essential elements of a valve quite unlike any valve I had seen before.'*

Other examples. These come from the authors' observations of engineering students' work on conceptual design tasks:

Firstly, ideas for automatically gripping and turning pages of books:

- *a suction device like a vacuum cleaner;*
- *a friction pad similar to a conductor's thumb removing tickets.*

Secondly, ideas about columns:

'The hollow cylinder is used in nature most effectively in reed stalks, bamboo shoots, and animal bones, and is about as strong under compressive load as a solid cylinder of comparable diameter.' (A statement made by young student with no knowledge of column theory.)

The chaos of human creativity is not readily reduced to the formality of the written word. De Bono (1976), in a series of anecdotal but insightful experiments, has drawn attention to the different levels of thinking and understanding adopted by people. The characteristic response to new and difficult problems (e.g. why did a vertical black cylinder standing upright on a table suddenly fall over during a de Bono talk?) is one of vague inchoate thinking drawing on poorly defined, fuzzy concepts; a sort of intellectual 'porridge' (the cylinder had an internal knocking over mechanism). Porridge words are valuable just because they are blurred and formless; they enable the thinker to keep options open and to leap-frog over areas of ignorance, and thus may be a prelude to successful problem solving (spring-loaded, retractable pin in base of cylinder, operated by solenoid and battery). Successful creative effort may thus require acceptance of the vague and imprecise and tolerance of the ambiguous.

The process by which Watt and Whittle made their inventions conforms to a pattern of discovery identified in other investigations of creativity in science and engineering (Hadamard, 1949; Wallas, 1926). The following sequence seems to be typical of highly creative individuals working on their own:

Recognition — The existence of an important and worthwhile problem is recognised.
Saturation — In an intensive attack on the problem the would-be problem solver saturates himself/herself in all its aspects without achieving a solution.
Incubation — The problem solver becomes interested in other matters; the problem ceases to occupy her/his conscious mind.
Illumination — Then in some chance moment of reflection, the problem solver's mind drifts back to the problem and an illuminating flash of inspiration is received — the 'Eureka' experience.
Elaboration — The idea is followed through and implemented.

If an inventive solution to a design problem is expected or demanded, then time should be allowed for incubation; for the problem to lie fallow in the problem solver's mind while attention is directed towards other matters. In addition, however, there are specific ways of stimulating creative effort, ways of raising our level of intellectual performance both as individuals and as members of groups in design teams.

2.3.3 Aids to creative effort

Common aids to creative thinking are listed and classified in Figure 2.5.

1. *Associative or combinatorial methods*
We may seek to stimulate creative effort by searching for new and useful combinations of existing elements. To apply this approach in a systematic manner presupposes that the system or device to be designed can be subdivided along two or more independent dimensions based on the functions to be performed and the components required to provide those functions. Then, each of the functional dimensions is brainstormed (see item 9 below) to compile as long a list as possible of ways of implementing it. Finally, a

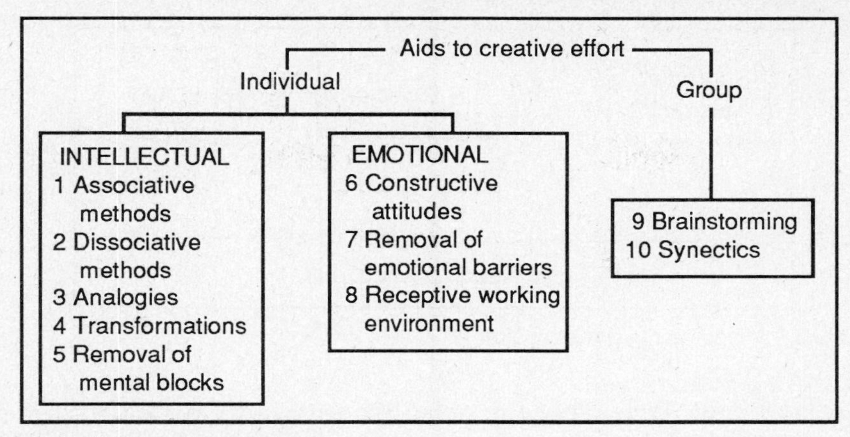

Figure 2.5 Aids to creative effort

matrix is drawn up showing functions versus means of implementation, so that all combinations are displayed for subsequent investigation. A couple of examples will help make this clear.

Example 2.e.3 *Drawing desk*

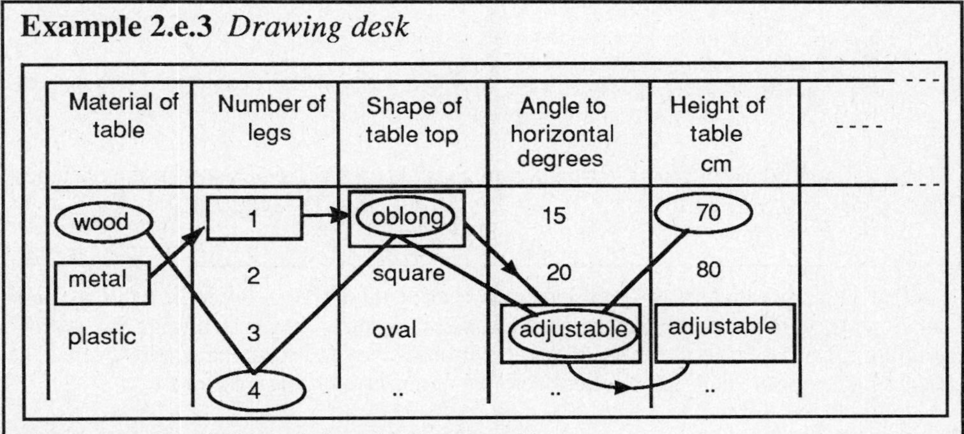

Figure 2.6 Morphological chart for drawing table

You are to design a new type of drawing desk to be used in an undergraduate design office. Use the morphological chart (Figure 2.6) to explore various alternative designs.

Using only the most commonly recognized elements of this problem, there are three material types listed (normally more would be used), four leg configurations, three table top shapes, three angles, and three table height variations shown in the chart. One could think of other elements of this problem, such as number of drawers and lighting arrangements for example. Thus the table in this sense is far from complete. Yet even in such basic form, the combinations available for consideration by the designer are:

$$3 \times 4 \times 3 \times 3 \times 3 = 324.$$

Two recognizable solutions are indicated.

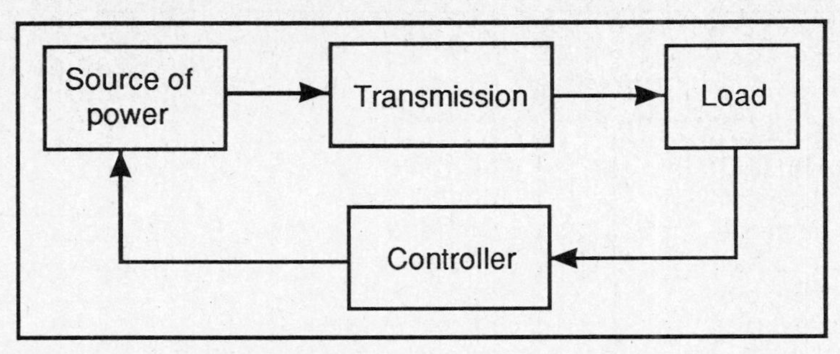

Figure 2.7 Engineering power system

Example 2.e.4 *Engineering power system*

Many engineering power systems are of the type shown in the block diagram (Figure 2.7), comprising:

1. load or device which absorbs power (e.g. the workpiece in a machine tool);
2. input source of power (e.g. electric motor);
3. transmission for matching power input to load (e.g. gearbox);
4. controls for starting, stopping, and maintaining stable operation.

Let N_1 = number of different output devices possible.
N_2 = number of different input devices possible.
N_3 = number of different transmissions possible.
N_4 = number of different controllers possible.

Then the number of different combinations available for consideration by the designer is:

$$N_1 \times N_2 \times N_3 \times N_4$$

The widespread use of combinatorial search methods is an indication of their inherent power, provided that in practice there is some means available for rapidly pruning the very large numbers of combinations generated to manageable levels. The following are some examples culled from the engineering design literature:

Bridgwater (1968): Combinations of unit processes in a systematic search for innovation in chemical engineering.
Wankel (1965): Alternative configurations of rotary engines.
Morgan (1971): Alternative layouts of machines and structures in power stations and for shoe manufacture.
Whitney (1987): Combinations of components for automated mechanical assemblies.

2. *Dissociative methods*

Applications of dissociative thinking to breaking existing, well-established linkages between the elements of a problem have been demonstrated in the cases of Watt and Cayley. To incorporate this mode of thinking as another weapon in one's intellectual armoury is frankly not easy. The process may be facilitated by the use of checklists of questions to be asked and reviews to be made in the search for new ideas. A well-known check list is exhibited in Figure 2.8.

Put to other uses?
New ways to use as is? Other uses if modified?

Adapt?
What else is like this? What other ideas does this suggest? Does past offer a parallel? What could I copy? Whom could I emulate?

Modify?
New twist? Change meaning, colour, motion, sound, odour, form, shape? Other changes?

Magnify?
What to add? More time? Greater frequency? Stronger? Higher? Longer? Thicker? Extra value? Plus ingredient? Duplicate? Multiply? Exaggerate?

Minify?
What to subtract? Smaller? Condensed? Miniature? Lower? Shorter? Lighter? Omit? Streamline? Split up? Understate?

Substitute?
Who else instead? What else instead? Other ingredient? Other material? Other process? Other power? Other place? Other approach? Other tone of voice?

Rearrange?
Interchange components? Other pattern? Other layout? Other sequence? Transpose cause and effect? Change place? Change schedule?

Reverse?
Transpose positive and negative? How about opposites? Turn it backward? Turn it upside down? Reverse roles? Change shoes? Turn tables? Turn other cheek?

Combine?
How about a blend, an alloy, an assortment, an ensemble? Combine units? Combine purposes? Combine appeals? Combine ideas?

Figure 2.8 Check list for new ideas
Source: Osborn, 1957.

3. *Use of analogies*
We have seen how the perception of analogies enlarges the mental resources of a problem solver by providing a richer variety of ideas as possible solutions to the problem. The role of analogies in major scientific discoveries has been discussed by Dreistadt (1968). A metaphor is an analogy expressed verbally. Schon (1967) sees the metaphor as a pervasive influence in language and fundamental to the creation of solutions to difficult, open-ended problems. Gordon (1961) has argued that thinking by analogy can be consciously used to stimulate new ideas. Solutions to particular problems can often be suggested by analogous situations in other problems, perhaps those found in the natural world of plants and animals. Nature has found so many ways of doing things that it provides a wealth of potentially useful new ideas. There are, for example, numerous different kinds of 'pumps' represented by the hearts of various animals.

4. *Transformations*
Transformations of problems are made with the intention of breaking the psychological set induced by previous experiences. *Set* is a concept used by psychologists to denote a

predisposition to a particular method or way of thought when solving a problem. 'Being in a rut' is a common expression with the same meaning. Some possible transformations are:

(a) changing the language in which the problem is expressed — if it is stated in words, turning it into mathematical or graphical form by means of symbols, flow charts, or networks;
(b) changing the boundaries of the problem — breaking implied constraints;
(c) inverting the problem — turning inputs into outputs and vice versa.

Set is not to be considered always bad just because it is sometimes an obstacle to new or inventive ideas. Set is the result of learning. Having learnt a particular method, one quite naturally wants to use it again. In breaking mental set, it is not necessary to say, '*Forget what is known*' or '*Never use an old method*', but rather, '*Remember there are many methods, not just one*'. It is not hard to correct for set once one is aware of its existence.

5. *Mental blocks*

Adams (1987) has written a spirited and entertaining account of how we make life difficult for ourselves by imposing blocks to our thinking processes. Are we too limited, too constrained in our thinking? Are we confused by masses of information and unable to see the wood for the trees? Do we see what we expect to see because our mental images are stereotyped? Are we so overwhelmed with information that we lose sight of essential detail? '*People talking without saying anything, people hearing without listening*', said Paul Simon and Art Garfunkel in their song, 'The Sounds of Silence'. Perhaps we should add, '*People looking without seeing*'.

Exercise

Sketch: (a) the front door of the house or apartment where you live;
(b) the dashboard of an automobile;
(c) a lecture theatre or laboratory you frequently work in.

Did you get all the details right? What did you miss?

Figure 2.9 shows some famous examples of drawings where things are not what they seem, at least from a quick, superficial inspection.

Overcoming mental blocks stemming from too-ready acceptance of unstated assumptions, stereotyping, and information overload requires a high level of self-awareness; an ability to stand back and examine one's own thinking processes. This can often be assisted by talking to other people. The simple act of explaining one's difficulties to another person clarifies thinking and gives fresh impetus to seeing a problem in a new light.

6. *Constructive attitudes to problem solving*

Many famous engineering problem solvers have exhibited great persistence in their endeavours, for example, Edison's experiments which led him to create the tungsten filament light bulb and Pilkington's work on the development of the float glass process. Constructive attitudes to problem solving can be manifest in this and many other ways — the capacity to postpone criticism and evaluation being a notable example. Criticism and fear of criticism are known to inhibit creative thinking. So the generation of ideas

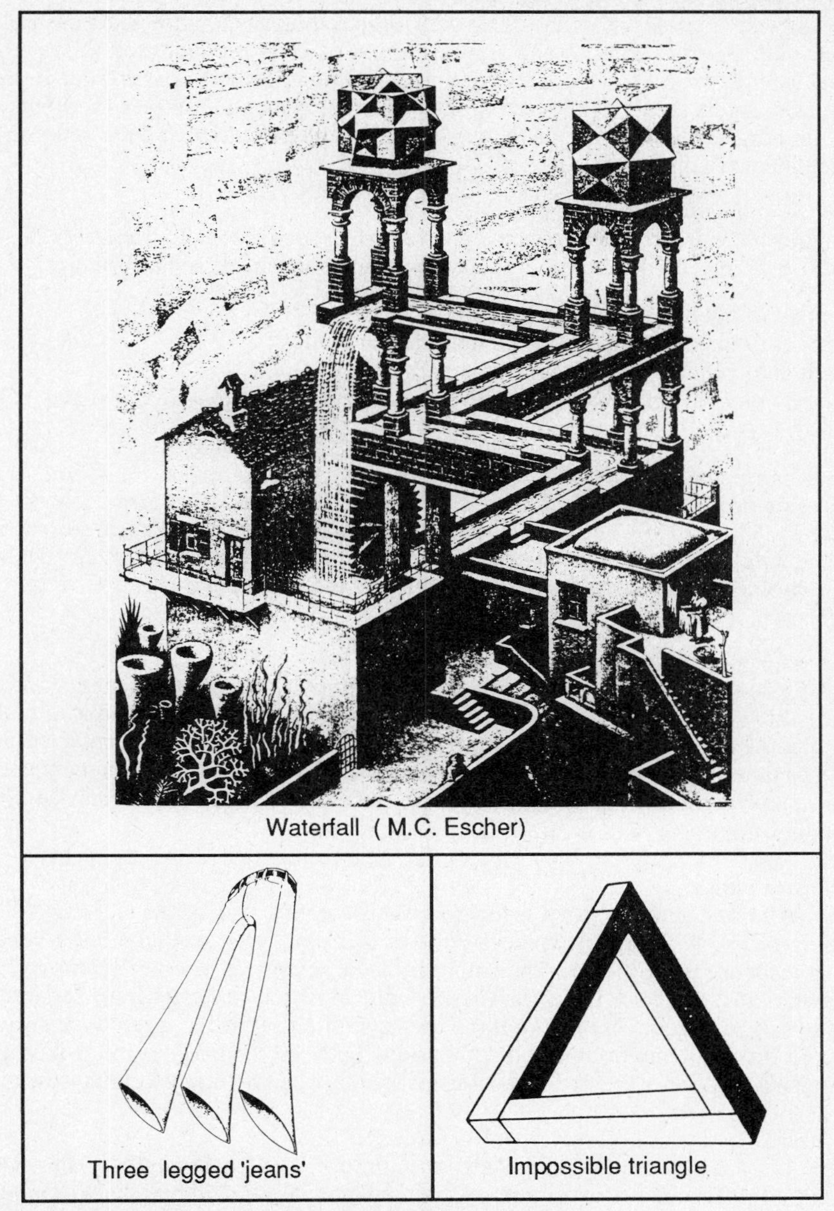

Figure 2.9 Drawings where things are not what they seem

for alternative proposals should be kept separate from the subsequent analysis of those proposals: creative effort gains from a free, uninhibited flow of ideas. Furthermore, we have seen how innovative designers are often presented with an amorphous fog of information, and are tolerant of this situation and of the need in the end to arrive at a precise engineering solution.

All too often designers, even experienced members of the profession, begrudge the time and effort needed to evaluate properly a new idea. De Bono's evidence (1977,

p. 90) suggests that as we grow older we become more jealous of the time we have invested in acquiring expertise and learning about existing products and processes. We are then less likely to keep an open mind and give new ideas the careful attention they deserve. Designers must be on guard against developing mental routines for prematurely ruling out new ideas and procedures, just because they are untried and the consequences of trying them are uncertain.

7. *Removal of emotional barriers*
An engineer's work may be adversely affected by emotional barriers which act as powerful restraints on the ability to think creatively. Some of these barriers are:

1. the fear of making mistakes;
2. the fear of exposing one's possible ignorance to others;
3. the fear of censure by colleague or superior; and
4. the fear of wasting valuable time invested in the development of an existing solution which is partially successful.

It is not surprising that research has shown that once one proposal for solving a novel engineering problem becomes preferred over any other by a design team, it is not easily rejected (Allen, 1966), even when major difficulties are encountered in its implementation. The recognition of possible emotional barriers is a necessary prelude to their removal. Their overthrow can stimulate the engineer's creative output very significantly.

8. *Receptive working environment*
The right working environment provides a receptive atmosphere filled with ideas, whereas the wrong environment can stifle the would-be innovator. Examples of adverse environments are: supervisors who measure results in terms of jobs completed and do not allow time for the hard thinking creative effort demands; colleagues unreceptive to new ideas, who conform to the old ways; the presence of noises and other disturbances which interrupt trains of thought.

9. *Brainstorming*
Probably the best known group technique for generating new ideas is 'brainstorming' (Osborn, 1957). A group of workers get together and offer an uninhibited stream of ideas in response to a problem. The emphasis is on *'letting the sparks fly'*; no evaluation is allowed, any criticism being postponed until after the brainstorming session. Like other group techniques, brainstorming requires skill and practice to produce successful results by breaking any prior emotional barriers between the participants. Dixon (1966) gives examples of how the free flow of ideas is stimulated by brainstorming sessions.

10. *Synectics*
Synectics is a group technique devised by Gordon (1961) and his associates. Gordon advocates the formation of groups containing people of different backgrounds and experience, and tries to promote creative effort by explicit use of analogies. He encourages empathy; the act of personally identifying oneself with whatever is being designed. The March 1968 issue of the *Journal of Engineering Education* includes a record of a synectics session.

2.3.4 Idea logs

Idea logs record in words and diagrams all the ideas conceived by a designer. The human brain moves from idea to idea very rapidly, and a facility for recording even the most fleeting thoughts is a great asset to the designer. A written record provides insurance against overlooking potentially valuable contributions; its contents may be a springboard for further bursts of creativity.

Extracts from a designer's notebook showing an idea log are displayed on the following pages. In this case the designer is an engineering student at the University of Melbourne who is tackling a problem concerning the dispensing of toothpaste.

2.4 Conclusion

In this chapter we have examined the divergent phase of the design process. It is now time to turn to the convergent phase in which the designer gradually converges onto the final solution. A central feature of this phase is decision making, the subject of the next chapter.

2.5 Notes from a designer's workbook

2.5.1 The 'bug' list

Most people react to day-to-day problems by the term 'this or that really upsets me', or 'it bugs me'. There is an amusing anecdote about bugs, dating back to faults in electrical equipment during the Second World War. Almost all electric controllers used electromagnetic relays. In the tropics some small insects would get in between the contactors of the relays and cause malfunction. Hence the phrase 'there is a bug in my equipment'. Nowadays we seem to refer to 'bugs' in computers or programs as an example of the organic nature of our language. Even we ourselves seem to be 'bugged' by various discontents surrounding us.

Example 2.e.5 *Preparing a bug list*

Prepare a list of things that disagree with you in your day-to-day environment. This list is best referred to as a 'bug' list. Try to stick to problems rather than solutions in your list. For example, *'I hate warm beer'* is a fair problem statement, but *'I wish I had a portable cooler for my beer'* is not acceptable.

Some other examples:

I hate getting up on cold mornings;
partly used toothpaste tubes are messy;
loud automobile exhausts are a nuisance;
my arthritic aunt can't open milk cartons;

2.5.2 An idea log

Idea logs are noted in Section 2.3.4. However a picture is worth a thousand words and the example below represents the type of idea log prepared early in the design to simply register various ideas on paper.

Example 2.e.6 *Idea log for toothpaste dispenser*

Tooth paste tubes become messy during use since there is no simple device available for dispensing cleanly from a tube. Prepare some sketch ideas for a dispenser capable of accommodating various types of toothpaste tubes and dispensing paste hygenically and cleanly.

IDEA LOG

PROBLEM

NEW TUBE.

TUBE USED UP WASTE—MESS

TOOTHPASTE DISPENSER

IDEA LOG.

* IMPROVE PRESENT METHOD WHICH RESULTS IN SCREWED UP TUBES AND UNTIDY LOOKING, WASTING TOOTHPASTE
* POSSIBILITIES
* DIFFERENT PACKAGING — SYRINGETYPE MECHANISM

PUSH

— MORE EXPENSIVE TO MANUFACTURE
— COULD BE REFILLABLE
— PROBLEMS WITH WASTAGE

IE. MORE COMING OUT THAN NEEDED

TOOTHPASTE WOULD GO STALE UNLESS A CAP OF SOME SORT WAS FITTED

Problem statement and problem development

PROBLEMS WITH WASTAGE OF PASTE NEAR FINISH OF TUBE, BECAUSE PLATES CANNOT COMPLETELY CLOSE.

— ROLLERS

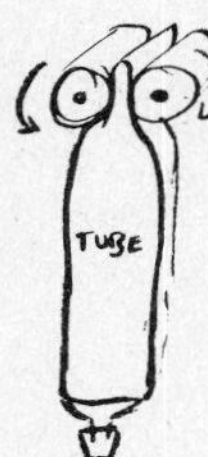

—NEED SOME CONVENIENT METHOD TO TURN ROLLERS, NOT EJECTING TOO MUCH PASTE

PLACE LEVER ON ONE ROLLER.

RATCHET TYPE CONNECTION B/W LEVER + ROLLER.

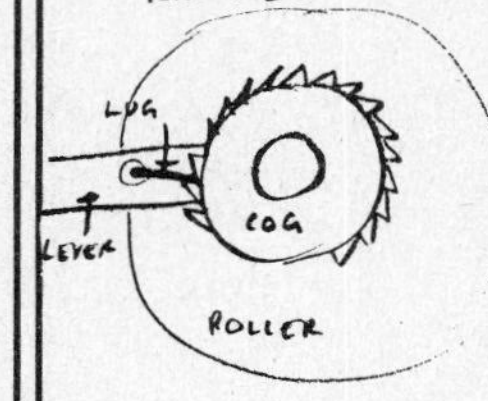

A LUG ON THE LEVER WHICH, WHEN LEVER MOVES UP, SLIDES OVER OF COG AND ON DOWN STROKE JAMS INTO THE TEETH + TURNS COG COULD BE USED.

ROLLERS MADE OF HARD RUBBER WOULD GRIP TUBE + WHEN PLACED VERY CLOSE TOGETHER, WOULD COMPLETELY SQUEEZE TUBE DRY.

THESE WOULD PROBABLY LOOK BETTER THAN METAL-FOIL TUBE, BUT MORE CLUMSY TO USE + IF DISPOSABLE WOULD WASTE MATERIALS (i.e. PLASTIC, METAL ETC.) IF NOT DISPOSABLE THEY WOULD NEED TO BE CLEANED REGULARLY. CLEANING DRYED TOOTHPASTE IS DIFFICULT & TIME CONSUMING.

* ADAPT PRESENT PACKAGING.

—WINDER ON END OF TUBE.

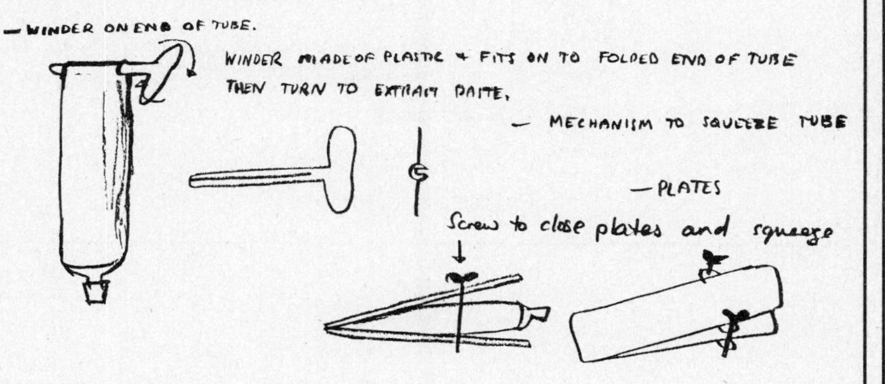

Further development of the problem

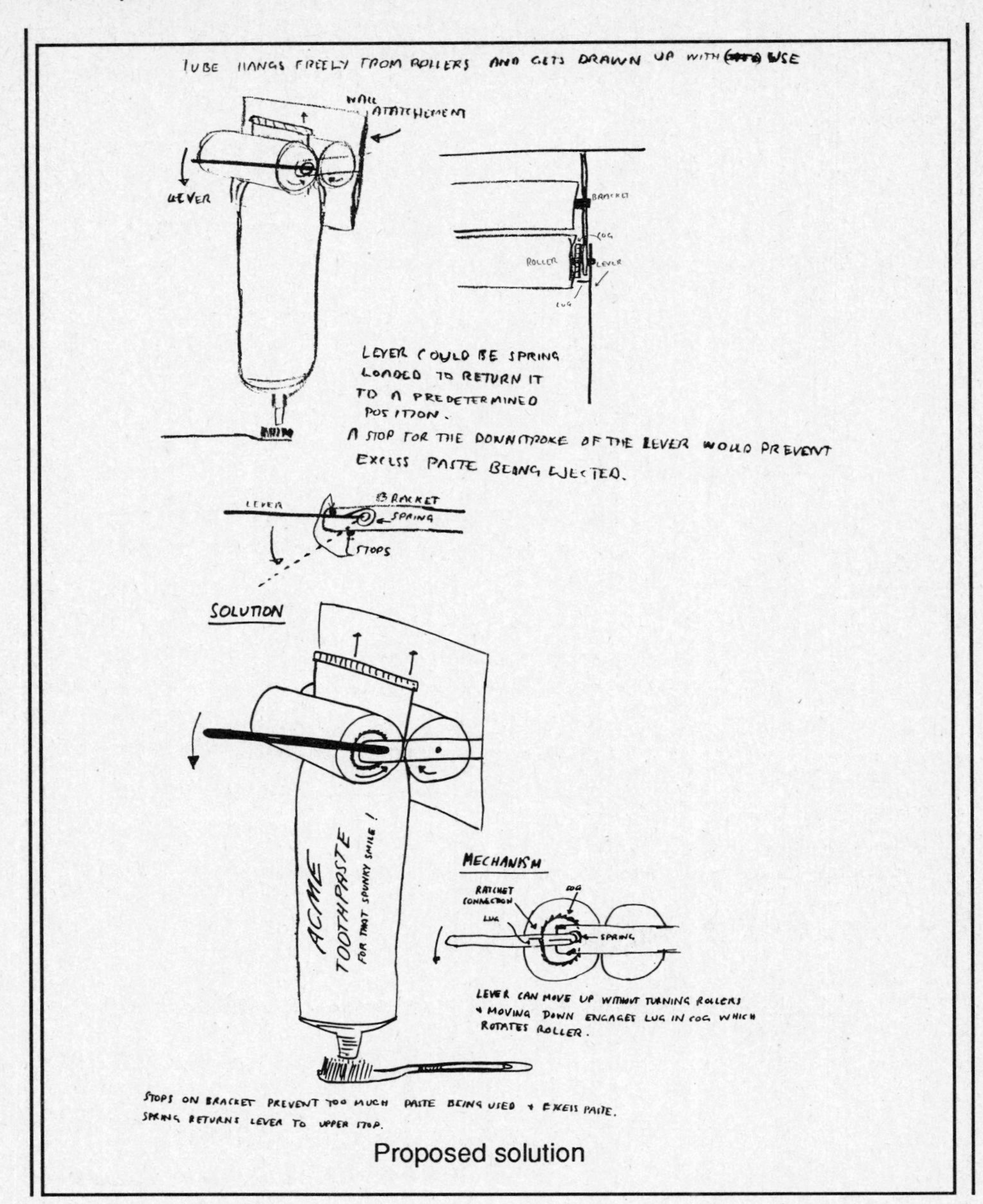

Proposed solution

Chapter 3

The convergent phase of the design process: Decision making

What is strength without a double share
of wisdom? — vast, unwieldy, burdensome,
Proudly secure, yet liable to fall
By weakest subtleties — not made to rule,
But to subserve where wisdom bears command!

John Milton

Concepts introduced decision making; probability; benefits and costs; strategy.

Methods presented decision tables; benefit-cost analysis; sequential decisions; trade-offs and compromises.

Applications (a) design of spur gear drive; (b) reliability of systems.

3.1 Introduction

The engineering designer uses various methods to investigate alternative proposals. Very often in practice, the iterations of synthesis and analysis follow one another so closely that these methods are bound up inextricably with the whole process of designing. Sometimes in special applications the analytical techniques can be reversed to synthesize a new design directly. As an example, consider two possible approaches to designing a wing profile (aerofoil section). Firstly, one can design a wing section to give a specified lift coefficient, and then analyze the distributions of air pressure and velocity around the wing to see whether they are acceptable from the point of view of drag and stall characteristics. If these characteristics are unsatisfactory, the original design would have to be modified and a new analysis made. By this process of successive trial and error one would converge onto the final solution. The second approach is to design the wing profile for specified distributions of pressure and velocity known to give rise to satisfactory aerodynamic performance. Techniques of 'direct design' for this purpose are available (with computational help from a computer). The wing profile so obtained would then have to be checked for structural effectiveness and suitability for manufacture. Iterations of synthesis and analysis would have to be made again.

In this chapter we are not concerned with special techniques of direct design, but rather with the general features of the convergent phase of the design process. A design proposal has first to be investigated to prove its feasibility. Its performance, and the associated benefits and costs, have then to be predicted so that its advantages and disadvantages can be compared with those of other proposals.

3.2 Feasibility studies

Objectives classed as essential are hurdles which any feasible proposal must be capable of surmounting. In general, feasibility studies are made to determine which of the various proposals put forward are feasible. The following questions are asked:

1. *Physical realizability.* Is the proposal in accordance with existing scientific knowledge? Can the hardware be made and assembled?
2. *Resources.* Are the financial and other resources available to implement the proposal?
3. *Economic worth.* Will the outcomes of the proposal be of sufficient value to repay the effort of implementing it?

Order-of-magnitude calculations are a common feature of feasibility studies to determine whether a proposal is in the right 'ball park'. A provisional decision may be made that a proposal is feasible although a final assessment may not be possible until more information has been generated during detailed design. Marples (1961) gives an example of this type of situation in the field of power station design.

3.3 Selecting one proposal from a number of alternatives

3.3.1 General

We consider the situation where there are several alternative courses of action open to the designer and a decision has to be made about which one to select.

A large number of factors will affect any engineering decision. A simple but useful way of formalizing the decision-making process is by means of a decision table such as that shown in Figure 3.1. This helps to ensure that all relevant factors are considered and that none are overlooked. The evidence from experimental psychology on the limited capacity of the human brain for processing information has been reviewed by Welford

Proposal ↓ / Criterion →	1	2	3	----
1	High	80%	3 years	
2	Medium	75%	7.5 years	
3	Low	50%	5 years	
⋮				

Figure 3.1 Decision table for systematic rating of proposals against criteria

(1971, Chapters 6 and 7). Because of this limit, a systematic method for comparing the advantages and disadvantages of alternative proposals is essential.

In the example in Figure 3.1 the criteria might be:

1. complexity of design;
2. efficiency of operation;
3. expected life.

Each cell in the decision table shows the rating of one proposal against one criterion. The construction of a decision table is relatively straightforward and objective. Difficulties arise when different criteria are measured on different dimensions using different scales, possibly even different types of scale (e.g. ordinal and interval). The resolution of these difficulties is further discussed in Section 3.5. In Section 3.3.2 below we examine decision making where:

1. the benefits and costs of alternative courses of action can each be expressed on one scale (not necessarily the same scale); and
2. the notion of probability can be introduced.

As an example of No. 1 and of the way in which vague and qualitative objectives are converted to quantitative criteria, see Middendorf (1969, p.37). He analyzes in detail a decision whether to use copper or aluminum cables for electricity transmission lines.

3.3.2 Probabilities

A course of action may have various consequences or outcomes which in this discussion are expressed as 'benefits' or 'costs'. In order to select one course of action from a number of possible alternatives the designer predicts the outcomes of each course of action and the probability that a particular outcome will follow a particular action. The designer estimates the desirability or undesirability of each outcome, and then decides on a course of action using some common scales for comparing the benefits and costs of the various outcomes. This description of the process of decision-making implies the following:

1. means of predicting the outcomes of the various courses of action open (i.e. simulation techniques);
2. predictions of the probabilities that particular courses of action will lead to particular outcomes;
3. sets of values for measuring the benefits and costs of outcomes;
4. a criterion for selecting the most desirable outcome.

When the above general description is applied to decision-making in engineering design certain features become apparent:

1. Engineering designers use a number of special modelling techniques as a means of predicting the outcomes of the alternative ideas under consideration. The most common techniques are :

   (a) mathematical;
   (b) graphical; and
   (c) physical scale models.

2. Engineering designers want to have a high degree of confidence in their predictions because the penalties of failure are severe. They will often make subsidiary decisions to give themselves sufficient confidence in the outcome of a critical decision, for example, they may decide to use a large factor of safety or to build and test a scale model of a new machine or plant.
3. Engineering designers try to quantify systems of values as far as they can, usually by costing the tangible outcomes of critical design decisions.

The contrast between innovative design and evolutionary design should be noted. In innovative design where there are few or no precedents, the emphasis is not on the prediction of the outcomes of alternative proposals in terms of performance objectives and costs but on the hurdle requirements of achieving the essential objectives. In the assessment of alternative solutions proposed for an innovative design problem the first lot of questions to be answered are:

- Will a proposal meet all essential objectives?
- Will it work?
- Is it feasible?

We now consider decision criteria.

Let: B_{ij} = benefit of jth outcome of ith proposal.
C_{ij} = cost of jth outcome i^{th} proposal.
p_{ij} = probability of obtaining benefit B_{ij}.
p'_{ij} = probability of incurring cost C_{ij}.

The expected total benefit and total cost of the ith proposal are:

$$B_i = \Sigma p_{ij} B_{ij}$$

$$C_i = \Sigma p'_{ij} C_{ij}$$

Common criteria for selecting i are based on expectation, as set out below:

1. C_i to be a minimum subject to $B_i > B_{min}$, where B_{min} is some prescribed minimum level of benefit;
2. B_i to be a maximum subject to $C_i < C_{max}$, where C_{max} is some prescribed maximum cost;
3. if B_i and C_i are expressed on the same scales, $(B_i - C_i)$ to be a maximum.

Another common criterion is based on level of risk: discard proposals whose p_{ij} are unacceptably low or whose p'_{ij} are unacceptably high. Other more specialized criteria are discussed in textbooks on operational research.

3.4 Strategies of decision making

3.4.1 Interacting decisions

Frequently the problem facing the designer is more complex than that described in Section 3.3, in that a decision has to be made to undertake several courses of action simultaneously, each action being selected from a number of possible alternatives and the outcomes of the different courses of action being interrelated. After an initial set of decisions has been taken, the interrelationship between the outcomes of these decisions may or may not be satisfactory. If unsatisfactory, some or all of the original decisions would have to be revised until their outcomes achieved the desired interrelationship.

Newcomers when faced with this situation feel confused and ill at ease, partly because they don't know where to start nor what initial set of decisions to make, and partly because of the heavy cognitive load they are working under. In an investigation of the design of powered artificial limbs, Wilson (1968) was led to comment,'*One can sometimes chase the problem like a will-o'-the-wisp on and on apparently indefinitely.*' He advocated the use of networks to show graphically which decisions directly interact with others because '*One cannot hold all the items in one's mind at once*'. For additional evidence of the disadvantages of relying on purely verbal ways of representing the interactions between decisions in a complex design problem, see Gott and Berridge's paper (1966) on power stations.

Rudd and Watson (1968) give a mathematical formulation of this state of affairs. They consider well-defined design problems which can be simulated mathematically, where the number of design variables whose numerical values are to be decided (M) is greater than the number of equations relating them (N), $M > N$. In general, the variables may be continuous or discrete and they may be subject to constraints of the form $a \leqslant x \leqslant b$. There are thus $F = M - N$ degrees of freedom. F of the variables may be selected as independent design variables x_d; the remaining dependent or state variables x_s are then fixed.

The design equations are of the following form:

$$f_i(x_{dj}, x_{sk}) = 0$$

where: $i = 1, 2, \ldots, N$ number of design equations.
$j = 1, 2, \ldots, F$ number of independent design variables.
$k = 1, 2, \ldots, N$ number of dependent design variables.

We have therefore to explore an F dimensional design space.

We look for a criterion function

$$C = c(x_{dj}, x_{sk})$$

which will enable us to determine an optimum design within the feasible region of this space. A two-dimensional example is shown in Figure 3.2, which is only approximately to scale. Other examples from different branches of engineering will be found in Rudd and Watson (chemical engineering), Start and Nicholls (civil engineering), Middendorf (electrical engineering), and Johnson (mechanical engineering).

3.4.2 Sequences of decisions in innovative design

In the assessment of alternative proposals for solving an innovative design problem the first questions asked are:

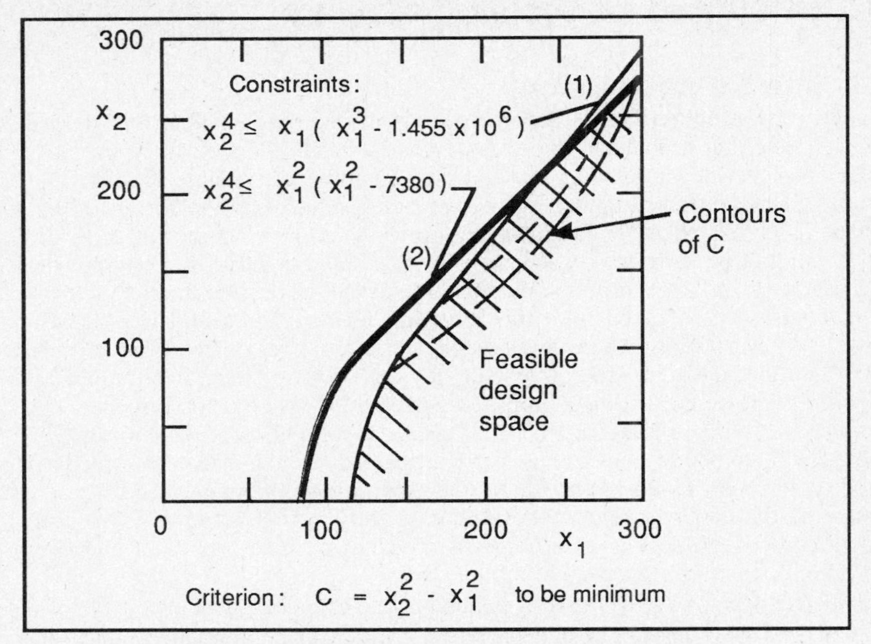

Figure 3.2 A simple example of a multidimensional design problem

- Will a proposal satisfy all essential objectives?
- Will it work?
- Is it feasible?

These questions can only be answered by identifying the subproblems thrown up by the proposal and estimating their tractability or ease of solution. Marples (1961) investigated a typical innovative design problem which concerned equipment for a new type of power station. He concluded that the sequence of problems and subproblems could be illustrated diagrammatically on a decision tree. The general case is shown in Figure 3.3, where the vertical lines represent the subproblems which have to be considered in the course of the design and the sloping lines are the alternative solutions proposed. The original design problem is designated p_o and the designer has decided that it can be broken down into a number of major problems $p_1, p_2, p_3, \ldots$ which are independent and can be tackled in parallel. a_{ij} is the jth alternative to subproblem p_i. p_{ijk} has in turn arisen from consideration of a_{ij}.

For example, if the problem given p_o is to design a new machine for sorting and grading citrus fruit according to size, this can be subdivided into three major parts:

1. the problem of conveying the fruit to the grading operation p_1;
2. the problem of devising a means of grading the fruit p_2;
3. the problem of delivering the graded fruit to the packers p_3.

Initially these three problems can be tackled independently with the designer needing to exercise creative ability to think up alternative ways, a_{ij}, of solving them.

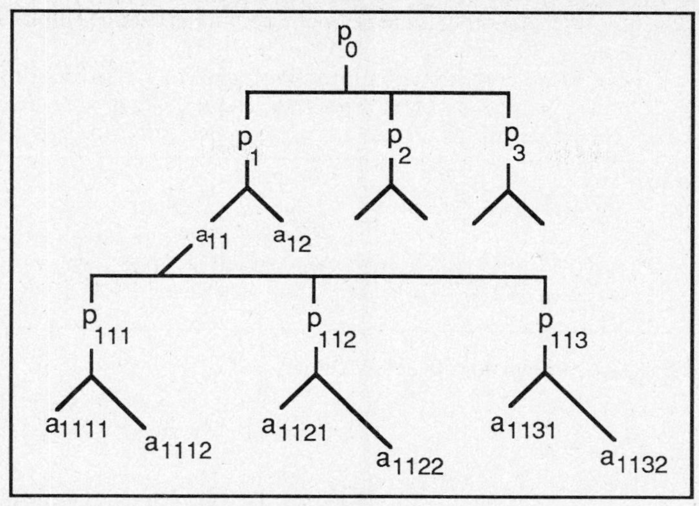

Figure 3.3 Decision tree for an innovative design

The designer then asks: 'Will these alternatives work?' This leads to investigation of subproblems P_{ijk} such as those of handling the fruit without bruising, and so on. To find out whether or not the subproblems can be solved, the possible alternative solutions a_{ijkl} must be examined.

To sum up, in innovative design problems the designer asks the following questions:

1. *Which of the proposed solutions are feasible*? Unworkable 'pseudo-solutions' must be discarded as soon as possible in order not to waste time on them. This leads to consideration of sub-problems according to a decision tree.
2. *Which of the feasible solutions is the best*? To answer this question, a decision criterion has to be applied — maximize performance or minimize cost , and so on, as discussed previously in Section 3.3.

3.4.3 Sequences of decisions in evolutionary design

In evolutionary design the science of engineering has had time to catch up with the art, and proven modelling techniques are therefore available from the application of scientific theory to guide the designer. Any doubts the designer might have in the predictions from these models may be removed by the incorporation of safety factors or other empirical coefficients, themselves justified by the successful performance of previous designs.

The decision tree or decision sequence diagram can be usefully extended to problems of evolutionary design. To illustrate, a case study on spur gear design will now be considered.

Example 3.e.1 *Spur gear drive*

The problem: Design a gearbox to transmit 25 kW. The input and output shafts are parallel, the input speed is 1,500 rpm, the output speed is 500 rpm, and the equivalent running time is six hours per day, a large number of gearboxes are to be made. It has already been decided to use spur gears with involute teeth and 20 degree pressure angle.

Criterion: The cost of the gearbox is to be a minimum consistent with its size, being less then 60 cm long by 20 cm wide by 37 cm high.

There is a very large number of alternative solutions representing possible combinations of different values of the design variables. Altogether there are nine design variables, the values of which are to be determined (refer to Figure 3.4).

Center distance	C	*Module*	m
Diameter of gear	d_g	*Material of pinion*	M_p
Face width	b	*Number of teeth in pinion*	T_p
Material of gear	M_g	*Number of teeth in gear*	T_g
Diameter of pinion	d_p		

Some of these quantities are related. We have:

$$d_p + d_g = 2C$$

$$\frac{d_p}{d_g} = \frac{1}{3}$$

$$m = \frac{d_p}{T_p} = \frac{d_g}{T_g}$$

There are also equations which express the strength of the teeth of the pinion and gear — their ability to resist bending and wear. When these various relationships are taken into account, the designer finds that there are three critical decisions to be made, and there is a definite logical order in which they should be made. In fact, center distance is first selected, then diametral pitch, and then the material of the pinion. It would be foolish and time-wasting to start the design calculations by choosing face width or the material of the gear.

Figure 3.4 shows the sequence of decisions in this case. On the right hand side of Figure 3.4 the new information that becomes available during the course of the design is tabulated. Firstly, a decision on center distance yields the pitch diameters and the tangential load at the pitch radius to transmit the required torque. Secondly, a decision on module gives us the numbers of teeth on the pinion and gear. It also gives information on tooth interference, contact ratio, and the strength and geometry factors used in American and British Standards. Thirdly, a decision on what material

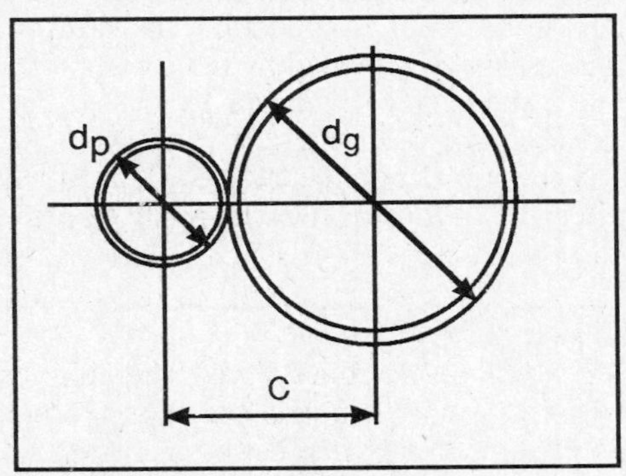

Figure 3.4 Geometry of a pair of spur gears

to use in the pinion gives rise directly to the face width and then the material of the gear follows. At this stage enough information has been generated for an estimate to be made of the overall dimensions of the gearbox and its cost. This can be done for various alternative combinations of center distance, module, and pinion material. The results are then compared and the decision criterion applied to determine the combination of design variables to be used.

Note that in a quantitative design such as this, the design variables are either continuous or discontinuous. In this case, center distance and face width are continuous, while the other variables can only take a limited number of discrete values.

The case study serves to illustrate that in engineering design we are often just as much concerned with the setting up of a logical sequence of decisions as we are with individual decisions — a sequence in which the information derived from one decision forms a rational basis for the next. Can we formulate rules for arriving at the best decision sequence, that is, the sequence in which the designer has finally to consider the minimum number of alternatives (i.e. the minimum number of alternative gearboxes in the above example)?

From an analysis of the case study of gearbox design, we deduce the following rules:

First rule: *Adopt the decision sequence in which the minimum number of critical decisions are made.*

Second rule: *Ensure that early decisions in the sequence yield a maximum of useful information for later decisions.*

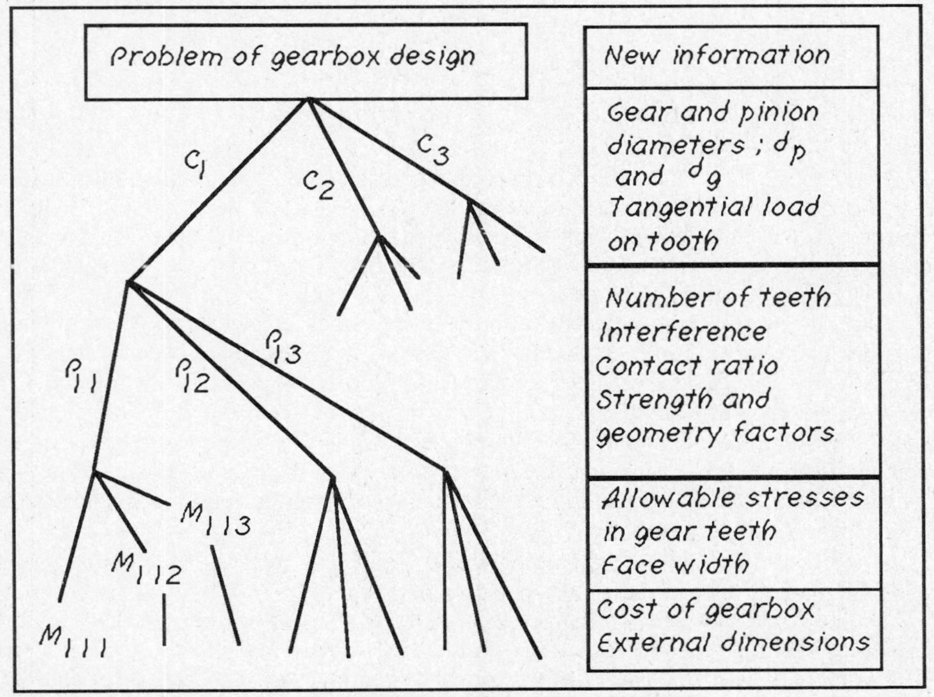

Figure 3.5 Decision sequence diagram — Spur gear design

In the case of the gearbox, the initial decision on center distance yielded the pitch diameters of the pinion and gear and the tangential tooth load — valuable information for later work. On the other hand, an initial decision on face width would not have yielded any other information at all.

Third rule: *Make critical decisions on discontinuous variables as early as possible in the design.*

In the case of the gearbox, design decisions on module and pinion material were restricted to a small finite number of possible alternatives, and these variables were therefore considered early in the design. In engineering design calculations discontinuities often arise from the limited number of materials available and from the limited number of sizes and shapes of these materials.

Further insight into the problem of designing the gearbox is obtained by considering the mathematical model relating the design variables. The governing equations can be expressed in the following way:

$$d_p + d_g = 2C \tag{3.1}$$

$$\frac{d_p}{d_g} = \frac{1}{3} \tag{3.2}$$

$$m = \frac{d_p}{Tp} \tag{3.3}$$

$$m = \frac{d_g}{T_g} \tag{3.4}$$

$$f_p\,(b, d_p, M_p, m, T_p, T_g\,) = 0 \tag{3.5}$$

$$f_p\,(b, d_g, M_g, m, T_g, T_p\,) = 0 \tag{3.6}$$

Equations (3.1) to (3.4) follow from simple geometric considerations. Equation (3.5) says that there is a relation between six of the design variables. It is in fact a complex relation and has to be found from the design equations, graphical data, and tabulated material properties in gear design standards. Equation (3.6) applies to the gear in the same way that (3.5) does to the pinion.

Since there are six relations involving nine variables, there are three critical decisions to be made. From equations (3.1) to (3.4) it is clear that C and m should be selected early because of the other variables that follow from them (d_p, d_g, T_p and T_g) and because m is discontinuous. M_p should appear early in the decision sequence because it is also discontinuous. C is chosen first because it effectively determines the volume of the gearbox which is needed for application of the overall decision criterion. Decisions on m and M_p then follow in the sequence described. Thus C, m and M_p are the independent design variables.

In contrast to the above, the Sydney Opera House is a well-known example of the results of making design decisions in a disorderly manner. In this instance an initial decision was made to use a roof structure of exciting shape. Decisions on the design of halls and auditoria to be placed underneath this roof structure then had to follow: which halls were to be used for opera, which for orchestral concerts, and which for chamber music? How were the audiences to be accommodated? Surely a logical sequence of critical decisions should have been:

1. What musical works are to be performed — Opera? Ballet? Symphonies? String quartets?
2. What size of audience is to be accommodated at each performance?

Analysis of this and other case studies reported in the literature suggests additional rules for selecting successful design strategies.

Fourth rule: *In the selection of independent design variables, preference should be given to those design variables which are most closely constrained (Rudd and Watson, 1968, p. 68).*

Fifth rule: *Use a decision sequence which leads to a small number of feedbacks of information from the outcomes of later decisions to the data input to earlier decisions. Any feedback loops should be short.*

With respect to the fifth rule, Lewis (1967) gives an example from the field of hydro-electric engineering where the design of new generating equipment had to be substantially modified as a result of information gained near the end of the first attempt at the design. Major backtracking like this should be avoided.

3.4.4 Computational strategies and mathematical modelling

Designers wish to predict the future performance of the artefacts they are designing. Their predictions are very often based on mathematical models. Even simple design problems generate quite complex sets of mathematical relations. The clue to the efficient solution of such problems is their *information flow structure*. The complexity of this structure depends on the number of design relations and the number of variables in those relations which are under the designer's control, that is, the (minimum) number of variables to be specified in order to solve the problem.

If V = number of design variables and E = number of equations relating these variables in the designer's mathematical model, then one can define the number of degrees of freedom (F) available to the designer as:

$$F = V - E$$

In general, $V > E$ and the designer has F independent decisions to make; once these decisions are made the solution to the problem has been determined. (In practice some relations may well be inequalities, but this introduces complications which are outside the scope of the present discussion.)

Suppose that in a particular case we have a mathematical model of the following form, the x's being the design variables:

$$f(x_1, x_2, x_4) = 0 \tag{3.7}$$
$$f(x_2, x_3, x_5) = 0 \tag{3.8}$$
$$f(x_1, x_5) = 0 \tag{3.9}$$

Then $V = 5$, $E = 3$ and $F = 2$. Alternative strategies are illustrated in the information flowgraphs in Figure 3.6 where the connecting lines represent items of information and the circles represent transformations of information.

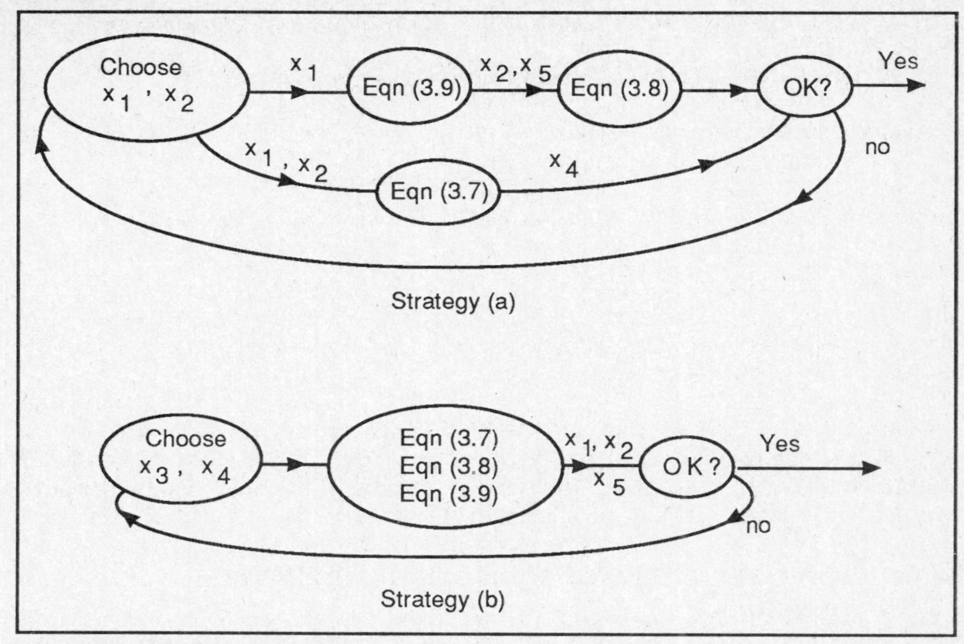

Figure 3.6 Information flowgraphs

Strategy (b) is inferior because it leads to a set of three simultaneous equations which in principle are much more difficult to solve than the single equations in strategy (a). Lee, Christensen and Rudd (1966) suggest the use of graphs of the form shown in Figure 3.7 to help determine superior design strategies. (See Harary (1969) for lucid exposition of graph theory.)

A design strategy which is represented by an acyclic graph is always superior to one represented by a cyclic graph.

3.4.5 The major aims of a decision strategy

In concluding this discussion we draw attention to the over-riding aims of any strategy of decisionmaking, namely:

1. to achieve a successful design with minimum expenditure of time and effort;
2. to ensure that all the interactions between the outcomes of different decisions are identified and taken into account.

3.5 Compromises

3.5.1 Introduction

We now consider decisions made under conflict. Suppose that one of a number of feasible alternative proposals is to be selected and that there are multiple objectives to be satisfied. With the ith objective is associated a criterion C_i measured on a scale S_i.

The simplest case occurs when there are two conflicting criteria and both can be measured on the same scale. Field (1970) discusses a design problem of this type in which there were two conflicting objectives, and after some manipulation each could be

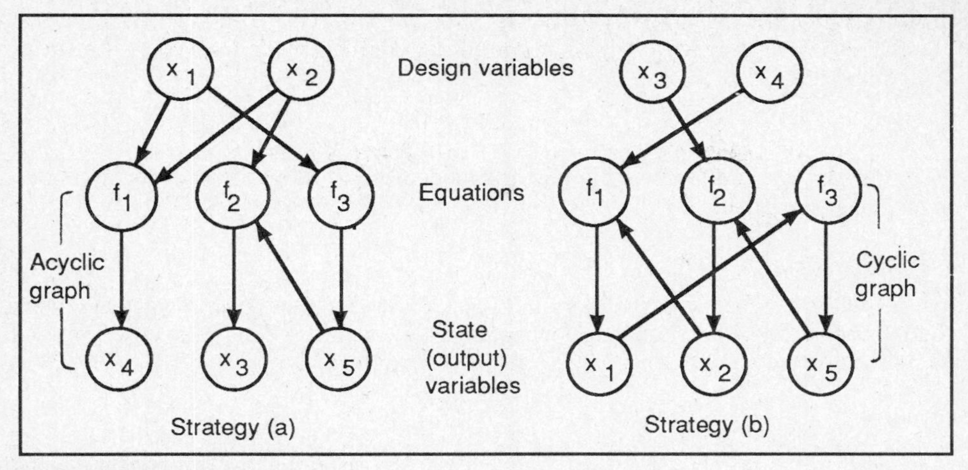

Figure 3.7 Strategy graphs

expressed on the same scale, namely cost in dollars.

The problem concerned the design of a heat exchanger, a piece of chemical process equipment whose function was to cool a liquid organic chemical flowing at a specified rate. Air was used as the cooling medium and increasing its velocity in the heat exchanger led to a higher rate of heat transfer from the process liquid, and hence to a smaller heat exchanger of lower capital cost. At the same time the pressure drop experienced by the air increased as did the power of the fan required to circulate it, so that running costs were higher. The compromise between the objectives of high rate of heat transfer and low air pressure drop was then made in such a way that the total cost of the heat exchanger (capital cost plus running cost) was a minimum.

A more complex state of affairs exists in design problems having two conflicting criteria which are measured on different dimensions. For example, Gasparini and Chong (1969) investigated the design of automobile side pillars — the two structural members supporting the roof on either side of the windscreen. Here the objective of structural integrity conflicts with that of minimum obstruction to the driver's vision of surrounding traffic. Note, however, that the scales used to measure the performance of a design against these criteria are both continuous and objective. Structural integrity is measured by the force in newtons required to break a side pillar, and obstruction to driver vision is measured by the angle subtended at the driver's eye by the region of obfuscation behind the pillar, in terms of number of degrees or radians.

Scales need not be continuous and comparisons may be subjective. Wehrli (1968) examined an architectural problem, the design of a primary school, and set up 24 criteria based on considerations of aesthetics, the social psychology of teaching, circulation of traffic in and around the school, and the use of the site in relation to its environment. The alternative designs proposed were rated on a scale of one to seven against each criterion by an expert judge.

Another example: A consumer magazine compared four electric razors all of which were on sale for approximately the same price. Each razor was given a rating on ten criteria by four judges who used each razor for a week. There were four ratings — poor to fair, fair, fair to good, good. The criteria included closeness of shave, speed of shaving, ability to shave over awkward contours, freedom from skin irritation, and so on.

3.5.2 Methods of compromise

We now review the principal methods used by engineers for resolving conflicts (Elmaghraby, 1968):

Dominant criterion

One criterion is of such overwhelming importance that it alone is used for decision making; others are ignored.

Composite criterion

A weight (w) is attached to each criterion in such a way that all criteria are mapped onto a common scale, for example, ratings on a scale of one to ten. A composite criterion can then be constructed:

$$C = \sum w_i C_i$$

Thus if the ith proposal has a score s_i on scale S_i, the best proposal will be that which maximizes:

$$\sum w_i s_i$$

Successive dominant criteria

In this approach one criterion is initially selected as the most important and all other criteria are ignored. Should more than one proposal have high performance against this criterion, attention is turned to another criterion — the most important one in the previously ignored set. Should more than one proposal rate highly against this second criterion, the next most important criterion is applied, and so on until one runs out of criteria or good proposals, whichever comes first.

This procedure is quite common in everyday life. For example, in the selection of a site for a new factory, one may specify the most important objective to be the availability of good road and rail transport. A number of possible sites may be equally superior in this respect; this is the set of 'best' proposals relative to the first criterion. Within this set one may now further search for a 'local best choice' relative to the availability of skilled labour and should there be a number of sites which are equally superior relative to this second criterion, one may then proceed to further refine the search for a 'best choice' within this set relative to tax and local financial incentives, and so on.

Dominant and threshold criteria

In this approach one criterion is selected as the most important (the primary criterion) and the others are expressed as constraints which alternative proposals must satisfy. This is done by setting the secondary criteria at their threshold values.

For example, suppose it is desired to design a control system for a space vehicle to the following specification. The system is composed of N stages which are connected in series. Stage i requires component i and for the sake of increased reliability several of these components are connected in parallel, say n_i components in parallel, (Figure 3.8). Assume each component i has a reliability (probability of non-failure) r_i, cost c_i, and weight w_i. It is desired to design the system of N stages which has the maximum reliability, minimum cost, and minimum weight.

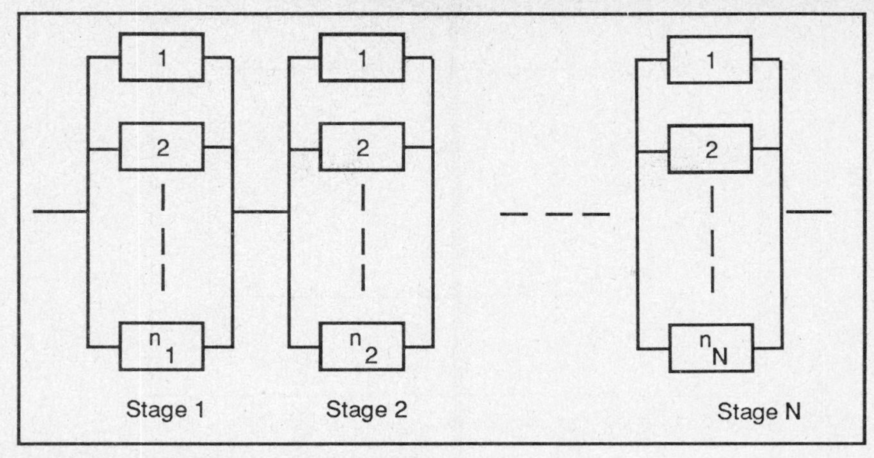

Figure 3.8 Block diagram of control system

The specification demands optimization relative to three criteria which are obviously contradictory. This is so since to maximize reliability one would increase redundancy by having many components in parallel at each stage, thereby increasing both cost and weight. Our approach to this design problem is to choose one of the three criteria as the primary criterion. Say we choose to minimize total cost:

$$C = \sum w_i \, C_i$$

and translate the other two criteria into constraints. For this purpose we require a level of reliability R considered to be satisfactory for the control system (say R is 0.99). We also require a maximum weight W which can be taken as an upper limit on the permissible weight of the control system. The constraints can then be expressed mathematically:

$$\prod_{i=1}^{N} \left\{ 1 - (1 - r_i)^{n_i} \right\} \geq R$$

$$\sum w_i \, n_i \leq W$$

The minimum cost configuration which satisfies these limits has then to be found. Of course, it is possible that the initial specification of the system was too stringent and that no solution can be found which satisfies both constraints simultaneously. In such a case either or both of the constraints would have to be relaxed.

Trade-offs and conflict curves

In the last example it would be helpful for the designer to know what percentage increase in weight (and cost) would correspond to a specified increase in reliability. The synthesis and analysis of alternative proposals would yield information of the form shown in Figure 3.9.

Figure 3.9 shows the possible trade-off between weight and reliability. If a reliability of 98 percent were to be accepted instead of 99 percent, the weight would be 7,000 units instead of 10,000, a reduction of 30 percent. In this hypothetical example a 1 percent

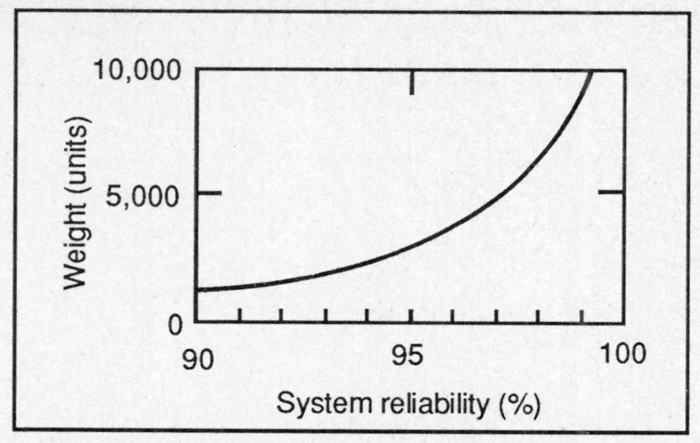

Figure 3.9 Variation of weight of control system with reliability

drop in reliability would reduce weight by 30 percent.

Conflicting criteria can be expressed graphically even when the scales are subjective. In the management of natural resources such as bushland some objectives might be:

1. maximum recreation use;
2. minimum disturbance to wildlife;
3. minimum water pollution.

Following McHarg (1970), one can estimate the form of the relationships between these criteria and attempt to express them graphically on subjective rating scales of 0 to 1. Thus, one would expect little conflict between 1 and 2 since whatever is done to improve water quality should also encourage wildlife. On the other hand, there would clearly be serious conflict between 1 and 3. Figure 3.10 has been constructed on the basis of reasoning such as this.

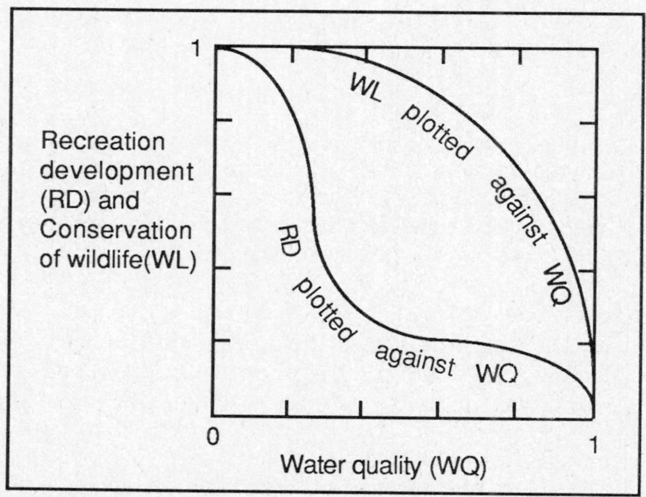

Figure 3.10 Conflicting objectives in land use

Example 3.e.2 *A difficult decision*

Suppose a person with a sweet tooth (let us call him or her X) is presented with an open box of chocolates by a hostess at a party and is invited to take one. Now X is bedazzled, stunned, confused. Each chocolate is a temptation, but only one can be chosen as the box will soon be on its way to the next guest. *'Mmmm,'* X ponders, *'the one with the hazelnut on top? Or perhaps the milk chocolate with the ripple pattern? Or the dark chocolate sprinkled in coconut? Or* 'Alas time is ticking by and a decision must be made'.

Now X is an engineer and recalls studies of decision making in engineering design. *'Having a sweet tooth, my dominant criterion for choosing a chocolate is size,'* X argues, and isolates the four biggest chocolates. *'Next, I like dark chocolate,'* and this narrows the choice to two. Then, *'The one on the left has a hard center and it will last longer.'* The decision made, X's hand stretches out and deftly selects the chosen confection as the hostess moves on.

Selecting a chocolate is not an act of much consequence. However, the method of selection is of general significance. The successive dominant criteria applied by our hero X were:

1. size of chocolate — the bigger the better;
2. colour — dark chocolate preferred to milk chocolate;
3. the length of time the chocolate was expected to last.

3.6 Conclusion

In this chapter we have been primarily concerned with methods of decision making. The successful application of these methods requires knowledge of the engineering sciences, economics (resource factors) and ergonomics (human factors). Knowledge of engineering science is gained from university courses in appropriate disciplines, and is outside the scope of this book. We now review some important concepts and methods in communications, economics, and human factors in Chapters 4, 5, and 6.

Chapter 4

Communicating formal messages: The design report

'Cheshire puss . . . ,,' she went on, "Would you tell me, please, which way I ought to go from here ?"
'That depends a good deal on where you want to get to," said the Cat.
'I don't much care where, " said Alice.
'Then it doesn't matter which way you go, ' said the Cat.
'So long as I get somewhere, ' Alice added, as an explanation.
'Oh you're sure to do that,' said the Cat, 'If you only walk long enough.'
Lewis Carroll, *Alice in Wonderland*

Concepts introduced — message; communication; writer and audience.

Methods presented — structure and format of reports; six key questions.

4.1 Introduction

Note: The Lord's Prayer comprises 56 words, the 23rd Psalm 118 words, the Gettysburg address 226 words and the Ten Commandments 297 words. By comparison, the US Department of Agriculture's order on the recommended price of cabbages has 15,629 words.

The design report is an essential component of most engineering projects, student and professional alike. It is the culmination of many hours of work and represents the sum total of current knowledge of a particular topic, leading to plans for alternative courses of action, then selection of the best plan and recommendations for its implementation. Reporting is to engineering as language is to communication, and competence in report writing is mandatory for the successful designer. The skillful selection and management of words and phrases is essential for conveying appropriate messages to colleagues, clients and the general public.

4.2 Some common errors and misconceptions

1. *Knowledge is power*

The formal structure of engineering education and the courses of instruction of which it is comprised induces a climate of opinion among student engineers which places a premium on the knowledge gained in these courses — usually in science-based subjects and mathematics where the acquisition of quanta of information is amenable to testing in examinations. Armed with knowledge of science and technology, the young engineer is ready for launching into a world of ageing button-pushers and paper-shufflers who, weary of turning the wheels of industry, are only too eager to pay homage to the young St Georges of the profession. It seems that all the old-timers have to do is point the way to the problem dragon and fade into the background.

This scenario, while stated in rather extreme terms, indicates the perception many young people have of the engineer as a person of action. What is overlooked is the need for co-operative effort from many persons to carry through to a successful conclusion any project of significance, and the need for facilitating that co-operation in order to make the best use of the human resources available. If you, the reader, believe that engineering is action not diluted nor contaminated by thinking, planning and communicating, then *read no further, this book is hazardous to your health and mental well-being.*

2. *Report writing, like wine, improves with age*

Drink makes a very dull man (paraphrasing Henry Fielding in *Tom Jones*), but only if he is a dull fellow to begin with. Quality wine improves with age, poor quality wine has nothing to recommend it. So too is the case with report writing. If the ingredients of good report writing are present, the engineer, as experience is gained, will become more skilled at mixing them together for the reader's delectation.

3. *Quality and quantity are transmutable*

On an occasion when admonished for verbosity, Sir Winston Churchill is credited with the comment, *'I don't seem to find time to write short reports.'* Short reports are a distillation of thought and skill with a dash of experience and flair for choosing the telling components of a story.

4. *It must be true: I read it in* The Times

The printed word commands authority. Its very existence derives from the fact that somebody has gone to the trouble of preparing a paper and imprinting words on it. Inevitably the reader assumes that all the work has not been in vain; that the words so presented have intrinsic significance.

There is a spectrum of credibility ranging from well-established facts (the speed of light is 300,000 km per second), through believable anecdotes based on personal interviews ('I slipped on the soap while having a cold shower this morning'), to wild generalizations (usually prefaced by propagandists with the phrase, 'As everyone knows...'). Designers, of necessity, plan for uncertain futures. It is their responsibility to delineate clearly the credence to be given to all statements in their reports. The reader must be given every opportunity to follow the thread of an argument, indeed to retrace the steps in the designer's thinking and lay bare the foundations on which it is constructed.

4.3 The ingredients of good report writing

Good engineering reports must provide answers to six basic questions:

Who? What? When? Where? Why? How?

1. *Who* ? Who is writing the report? Who is the audience to whom the report is addressed?

It is good practice to provide answers to the 'Who' questions on the front cover or title page of engineering reports.

2. *What*? What is the report about? What were the features of the design problem? What were the goals of the work? What was done to reach the goals? What conclusions have been drawn?

The answers to the 'What' questions are provided in several ways. A statement of goals early in the report answers one important question. An executive summary early in the report with a short review of the main achievements answers many of the other 'Whats' succinctly for the benefit of the busy reader. The body of the report sets out the answers to the question, 'What was done?' An engineering report will end with a statement of the conclusions reached, plus recommendations for any further work considered necessary.

3. *When*? and *Where*? These questions refer to the chronology and topography associated with the engineering work being reported. In general the answers are contained in the body of the report. However, a deeper interpretation of 'When' and 'Where' leads to detailed consideration of the contents of the report. The writer should continually ask questions such as 'When do I write about this aspect of the work or mention that reference?' 'Where does this piece of evidence best fit into my argument?' and 'Where do I put this theoretical analysis ?'

4. *Why*? Asking why elicits more specific questions, for example: Why did I do the work at all? Why this way? Why at this time? Why at this place? Why these decisions? Why these conclusions? A significant aspect of this process of question and answer is the continual self-examination and critical review of the writer's thinking that it entails.

5. *How*? How was the work done? How could it have been improved?

The reader of an engineering report should not be denied the benefits of hindsight.

4.4 A recipe for success

In a recipe for success, a sufficient, though not necessary, set of ingredients is presented together with a recipe for mixing them together.

Ingredients

1 title (well matured if possible)
1 quill of authors (usually one, two or three healthy specimens)
1 or more goals (well defined and firm)
1 summary (meaty and well formed)
1 body (with chunky sections on analysis, data gathering procedures, conclusions and other relevant information)
1 set of properly styled references
1 or more appendices

Preparation

Compose and collect ingredients and when suitably matured arrange in order. Season well with diagrams, sketches and other visual evidence as well as citations of references. Decorate to taste with binding and margins, 20 or 25 mm as desired on all pages.

Citation of references

References may be cited in the text of the report in several equally useful ways, as the following examples show:

1. Rubenstein (1952) — Identifies Rubenstein's 1952 publication in an alphabetically ordered list of references at the end of the report.
2. (Ref. 17) — References are listed in order of appearance in text; this is the seventeenth.
3. 'The prod fusillator was bent experimentally through an angle of 90 degrees [52]— Where the superscript identifies the 52nd reference listed.
4. 'When the fuel consumption reached 37 liters per kilometer, Clancy* lowered the boom' — Where the asterisk identifies a footnote at the bottom of the same page.

The same method of citation should be used consistently for all references.

4.5 On the lighter side

Not all report writing has to be deadly serious, and even in serious reports a light humorous touch can add welcome variety to a presentation and help maintain the reader's interest and good will. The examples which follow are taken from R. L. Weber's compilation, *A Random Walk in Science.*

The glossary for research reports illustrates in a light-hearted manner how the selection of key words and phrases can distort meaning as the writer strives for a positive self-image.

The British article, 'Which units of length?' (below), applies the procedures of rational decision making and evaluation against agreed criteria in an unusual context. The whimsical tone in which it is written helps to convey the message to the reader.

Glossary for research reports

(C. D. Graham (Jr), 1957) [1]

It has long been known that ...	I haven't bothered to look up the original reference.
... of great theoretical and practical importance.	... interesting to me.
While it has not been possible to provide definite answers to these questions ...	The experiment didn't work out, but I figured I could at least get a publication out of it.
The W-Pb system was chosen as especially suitable to show the predicted behaviour ...	The fellow in the next lab had some already made up.
High purity ... Very high purity ...	Composition unknown except for the exaggerated claims of the supplier.

[1] This set of phrases was compiled by C. D. Graham (Jr) in *Metal Progress,* Vol. 71, No. 75, 1957, and reported in R. L. Weber, *A Random Walk in Science*, The Institute of Physics, London, 1973.

A fiducial reference line ...	A scratch
Three of the samples were chosen for detailed study...	The results on the others didn't make sense and were ignored.
.... accidentally strained during mounting.	... dropped on the floor.
... handled with extreme care during the experiments.	.... not dropped on the floor.
Typical results are shown...	The best results are shown...
Although some detail has been lost in reproduction, it is clear from the original micrograph that ...	It is impossible to tell from the micrograph ...
Presumably at longer times ...	I didn't take time to find out.
The agreement with the predicted curve is excellent.	Fair
Good	Poor
Satisfactory	Doubtful
Fair	Imaginary
... as good as could be expected	[illegible]
These results will be reported at a later date	I might possibly get around to this sometime.
The most reliable values are those of Jones	He was a student of mine.
It is suggested that ... It is believed that ... It may be that ...	I think
It is generally argued that ...	A couple of other guys think so too.
It might be argued that ...	I have such a good answer to this objection that I shall now raise it.
It is clear that much additional work will be required before a complete understanding ...	I don't understand it.
Unfortunately a quantitative theory to account for these effects has not been formulated.	Neither does anybody else.
Correct within an order of magnitude	Wrong

It is hoped that this work will stimulate further work in the field.	This paper isn't very good, but neither are any of the others in this miserable subject.
Thanks are due to Joe Glotz for assistance with the experiments and to John Doe for valuable discussions.	Glotz did the work and Doe explained what it meant.

Which units of length?

(Pamela Anderton, 1973)[2]

Units of length have been available to the general public for a long time but the recent drive to advertise one particular brand has led us to publish this report for the assistance of members.

Brands

We found that the units fell into fairly well defined brands or 'systems' from which we have selected three in general use. Two of these, the 'Rule of Thumb' and the 'British' (known as 'Imperial Standard' in the days when we had an empire), are manufactured in this country; the third, the 'Metric', is imported but fairly readily obtainable.

Tests

We asked a panel of members to use units of the selected brands and to comment on their convenience. We also submitted samples to a well-known laboratory to find out how reliable they were. The selected units and the results of the tests are listed in the table.

Brand	*Unit*	*Reliability*	*Convenience in use*
Metric	micron	excellent	fair[1]
British	thou	good	good
Rule of Thumb	hair's breadth	poor	hopeless
Metric	millimeter	excellent	fair[2]
British	inch	good	good
Rule of Thumb	thumb	poor	excellent
Metric	meter	excellent	good
British	yard	fair to good[3]	good
	foot	good	good
Rule of Thumb	pace of stride	fair	excellent
	foot (i.e. size of shoe)	fair to good[4]	excellent

[1] Difficult to handle for everyday use and available to special order only.

[2] Our panel found it is about 25.4 times too small.

[3] Some samples tended to shrink.

[4] Users with big feet get better results.

[2] In R. L. Weber, *A Random Walk in Science*, The Institute of Physics, London, 1973.

Conclusions

The 'Rule of Thumb' was cheap, robust, very convenient and readily obtainable. On the other hand, it was not sufficiently accurate for all purposes.

The 'British' was convenient and readily obtainable, but some doubts exist as to its reliability. Nevertheless it seems likely to remain popular for a long time.

The 'Metric' is very reliable but not always as convenient to use as other brands.

Best Buys

For general use — Rule of Thumb.

For scientists and for others whose arithmetic is weak — Metric.

Chapter 5

Economic factors in design

Money is like a sixth sense without which you cannot make a complete use of the other five.

W. Somerset Maugham

Concepts introduced	present value, return on investment.
Methods presented	discounted cash flows, life cycle costing.
Application	economic evaluation of a new technology in vehicle design.

5.1 Introduction

Designers are responsible for specifying the ways in which engineering systems and products are to be constructed and manufactured. Many decisions taken by designers lead to the commitment of substantial resources by those responsible for the downstream activities of construction and manufacture. There is a heavy burden of responsibility on designers to ensure that these resources — time, effort and money — are applied in an effective and efficient manner.

In this chapter we are concerned with ways of measuring the value of resources committed to an engineering project and thus with the study of economic factors in engineering design. We examine the cost elements in manufacturing and then consider the time value of money, economic evaluation of projects and life cycle costs of products and components.

5.2 Cost to manufacture

A breakdown of the costs incurred by a typical manufacturing company is shown in Figure 5.1. The calculation of the return on their investment to the owners of the company is also set out.

Companies engaged in the batch manufacture of families of related products are often able to analyze their cost elements and express their direct and indirect manufacturing costs as functions of leading or characteristic product dimensions, typically by a simple power law such as:

$$C = a + bD^n$$

where:
C = element of direct or indirect manufacturing cost.
D = characteristic dimension.
a, b, n = empirical constants.

Examples of characteristic dimensions are:

- bore and stroke of engines;
- rotor diameter and width in turbomachinery;
- gear diameter and width in gearboxes.

Given the availability of cost information in this form, the designer can balance product performance against manufacturing cost, for example, to determine the optimum number of models required to cover the full range of performance, as in the case presented by McAree (1972) for pumping equipment.

Net profit before tax: = (Total sales) – (Total costs)

Working capital: = Accounts receivable
+ Cash in hand
+ Stocks
+ Work in progress
– Current liabilities

Total investment: = Fixed investment
+ Working capital

$$\text{Return on investment} = \frac{\text{Net profit}}{\text{Total investment}}$$

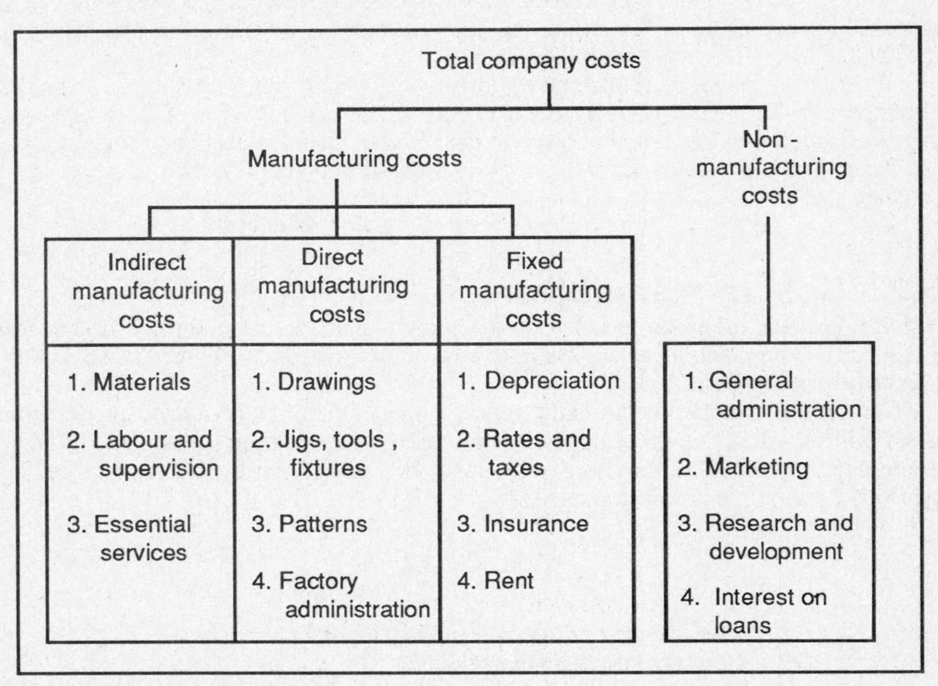

Figure 5.1 Total company costs

5.3 Economic evaluation of projects

In economic evaluations of new projects it is assumed that the benefits to be derived over their lifetimes and the costs incurred to provide those benefits can be expressed tangibly in terms of dollars. This is not to deny that other evaluations may be made, based on intangibles, concerned perhaps with aesthetics or the environment, which are difficult or impossible to measure. However, it is necessary to restrict the scope of the discussion to economics at this stage.

A given sum of money is worth more now than the same sum at a future date. Which would you prefer to receive as a gift: $1,000 now or the same sum adjusted for inflation in a year's time? It is necessary to find a method for comparing the dollar value of resources used at one time to the dollar value of resources at other times. This is done by computing the present value of these resources. The following example should make the procedure clear.

Example 5.e.1 *Present worth at different rates of interest*

Suppose an investor has *$1,000* to invest and three alternatives (*A*, *B*, and *C*) are available, producing the net cash returns tabulated, all figures being adjusted for the expected rates of inflation.

Table 5.1 Return on investment of $1,000

Year	*A* $	*B* $	*C* $
1	250	500	200
2	250	125	200
3	250	125	200
4	250	100	200
5	200	100	200
6	-	100	200
7	-	100	200
8	-	100	-
9	-	100	-
10	-	100	-

If the investor seeks a minimum return of 7 percent per annum, then the present values of the returns from the three alternatives are as follows:

$$A \quad \frac{250}{1.07} + \frac{250}{1.07^2} + \frac{250}{1.07^3} + \frac{200}{1.07^4} + \frac{250}{1.07^5} = 989$$

$$B \quad \frac{500}{1.07} + \frac{125}{1.07^2} + \frac{125}{1.07^3} + \frac{100}{1.07^4} + \ldots + \frac{100}{1.07^{10}} = 1{,}117$$

$$C \quad \frac{200}{1.07} + \frac{200}{1.07^2} + \frac{200}{1.07^3} + \ldots + \frac{200}{1.07^7} = 1{,}078$$

The order in which the alternatives are ranked is therefore *B*, *C*, *A*. In fact *A* is not acceptable at all since the present value of the net cash returns over the life of the investment is less than the initial capital outlay of $ 1,000.

In general, if P dollars are invested now for a period of n years at an interest rate of r per annum ($r = 0.07$ corresponds to an interest rate of 7 percent), then at the end of the n years the sum invested has increased to S where:

$$S = P(1+r)^n$$

Inverting this argument, we describe P as the present value of S, n years into the future. The ratio P/S is the present worth factor, *p.w.f.*, so that:

$$p.w.f. = \frac{1}{(1+r)^n}$$

Situations arise where a stream of equal payments, each R dollars say, is made annually for a period of n years. To find the present value P of this annual income (or expenditure) we note that the present value of the first payment is $R/(1+r)$, that of the second payment is $R/(1+r)^2$, and so on, where once again r is the prevailing interest rate. Hence:

$$P = \frac{R}{1+r} + \frac{R}{(1+r)^2} + \frac{R}{(1+r)^3} + \ldots\ldots + \frac{R}{(1+r)^n}$$

$$P = \left\{\frac{(1+r)^n - 1}{r(1+r)^n}\right\} R$$

The ratio R/P is called the capital recovery factor, *c.r.f.*, so that:

$$c.r.f. = \left\{\frac{r(1+r)^n}{(1+r)^n - 1}\right\}$$

The capital recovery factor (R/P) is that proportion of an initial investment P in a project that has to be returned as benefit R in each of the n years of the project's life to give the same value as the sum invested.

The apparently straightforward nature of the above results is deceptive. The value of r has to be discounted for inflation, and both interest rates and inflation rates may vary significantly with time. In practice, economic evaluations may require consideration of several different scenarios based on a range of possible future interest and inflation rates. Consideration may also have to be given to opportunity costs, because capital for investment is in short supply. Capital used for one purpose cannot be used for another productive purpose whose benefits are thereby foregone. The net benefit foregone is referred to as the *opportunity cost* of the capital. One can extend this line of reasoning to other scarce resources such as time. What would you, the reader, estimate to be the *opportunity cost* of the time you travel to work, in dollars per hour?

5.4 Life cycle costs

The consumer, user, or operator of an engineering product — whatever sort of system or device it is — has to predict and evaluate the costs to be incurred during its operating life, that is, the life cycle costs comprising:

1. *first cost* of purchasing the product and installing it; and
2. *annual cost* of using the product, consisting of:
   (a) fixed costs — those costs associated with the use of the product which will remain relatively constant throughout its life (e.g, interest and insurance charges, depreciation);
   (b) variable costs — those costs associated with the use of a product which will vary in some relationship with the level of that use (e.g. direct operating costs — labour, fuel, power, and maintenance costs). During scheduled maintenance there will be costs of repairs and replacements. During unscheduled maintenance (due to breakdowns) there will be additional costs incurred due to interruption of other activities (e.g. interruptions to production processes).

Depreciation is the decrease in value of the product with time. This decrease may be due to physical wear and tear or to obsolescence arising from technological progress. When considering depreciation as a cost it is necessary to decide at what rate the value of the product will decrease and to estimate the salvage value at the end of its service life. Suppose a product has a first cost of $5,000, will be used for five years, and is estimated to have a salvage value of $1,000 at the end of that time. If the value of the product is assumed to decrease at a constant rate, the depreciation each year will be $800. The rate of depreciation applied in economic assessments of equipment will also depend on national tax legislation and the rates allowed under that legislation for calculating the net income of the taxpayer who owns the equipment. In practice, depreciation models other than the simple straight line one may be used. A difficult decision may have to be made as to whether depreciation should be calculated on historic first cost or on replacement cost, that is, the cost of replacing the product at the end of its service life.

Example 5.e.2 *Life cycle costing*

Fuel conservation strategies for road vehicles powered by internal combustion engines are sensitive to changes in vehicle and fuel technology and to social and political actions affecting demand. In this example we look at the economic impact of some of the technological options in liquid fuel conservation. In particular we examine matters relating to vehicle design and apply arguments originally advanced by Watson and Milkins(1981).

In evaluating the cost to the consumer of any new design technology on a present value basis, costs accruing in future years are brought back to equivalent present costs by discounting at the appropriate rate. The life cycle cost of an automobile can then be expressed in terms of the total cost of vehicle ownership per year (T dollars):

$$T = C + A + K\,(M + F\,N)$$

where:
- C = annual equivalent of capital invested to purchase the vehicle.
- A = fixed annual cost — registration, insurance, depreciation, C and A measured in dollars.
- F = fuel consumption, liters/km.
- K = distance travelled per year, km.
- M = distance dependent maintenance cost, dollars per kilometer.
- N = fuel price, dollars per liter.

It has been shown by Watson and Milkins(1981) and others that $C = 0.15\ P$, where P is the purchase price of the vehicle. It is reasonable to assume that A and M are independent of fuel conservation technologies, so that the incremental cost ΔT of changes to the remaining variables is given by:

$$\boxed{\Delta T = 0.15\ \Delta P + K\ \Delta(F x N)}$$

This equation provides a useful basis for reviewing the costs of fuel conservation technologies and their interactions with fuel price.

Suppose a particular technology leads to an increase in vehicle price of ΔP and a reduction in fuel consumption of f percent for no increase in total cost, that is, $\Delta T = 0$ at the break-even point. Then the ratio $\Delta P/f$ gives a measure of the upper bound for the cost of an acceptable new technology.

Tabulated below are the values of ΔP calculated in this way for a 1 percent reduction in fuel consumption, that is, $f = 1$. These calculations of the breakeven cost of a new technology have been made for $K = 15{,}000$ km, $F = 11.3, 9.0, 8.0$ liters per 100 km, $N = 0.35, 0.45$, and 0.55 dollars per liter.

Table 5.2 Calculation of values of ΔP

F ↓ \ N →	0.35	0.45	0.55
11.3	\$39.5	\$50.9	\$62.1
9.0	\$31.5	\$40.5	\$52.1
8.0	\$28.0	\$36.0	\$44.0

Thus, if fuel costs 0.55 dollars per liter, and vehicle fuel consumption is 9.0 liter/100 km, then a new technology giving a 5 percent reduction in fuel consumption should not add more than 5 x (\$52.1), say \$260, to the purchase price of a new vehicle.

Chapter 6

Ergonomics

God is really only another artist. He invented the giraffe, the elephant and the cat. He just goes on trying other things.

Pablo Picasso

Concepts introduced	anthropometry, human information processing, systems and control, personal characteristics, safety.
Methods presented	closed-loop systems, time history of accidents.
Application	design of work area for engineering students.

6.1 Introduction

In this chapter we review what is known about engineering systems where people play a significant role in achieving successful performance.

Interest in human performance was stimulated by the Second World War. In the 1940s the technical development of high-speed aircraft, radar and other devices reached a point at which the limitations of people and machines working together were no longer mainly in the machines but in the human operators. Since then engineers have come to realize the necessity of considering the potentialities and limitations of the human operator if the equipment they design is to be used to the best advantage. In parallel with this, psychologists have set new criteria of adequacy in design which, although not always convenient for the engineer, are clearly important. The lessons learnt are now being applied to the design of workspaces, machinery, cars, household equipment, and so on.

Figure 6.1 (from Murrell, 1965) presents an overview of the subject. It shows the relationship between the inputs to and the outputs from a person acting as an element in a closed loop system, and sets out the critical factors affecting the person's performance in that system.

Human behaviour is extremely complex. Designers use a number of different conceptual models to help order their thinking about the different aspects of human behaviour which affect the equipment they are designing. These are:

1. occupant of a workspace;
2. source of power;
3. sensor or transducer;
4. processor of information;
5. tracker and controller;
6. person with motives, emotions, habits, preferences, maybe even some prejudices.

The remainder of this chapter is devoted to a consideration of ergonomics (or human factors) under the above headings. Important matters concerning aesthetics and safety are also discussed.

To conclude this introduction we summarize the factors affecting human performance at a particular task:

1. individual differences;
2. the nature of the task;
3. the training received for the task;
4. the environment:
   (a) physical;
   (b) behavioural:
      (i) formal: rules, laws;
      (ii) informal: customs, social mores;
5. The motivation:
   (a) intrinsic;
   (b) extrinsic.

For further reading, the following references will be helpful: Morgan et al. (1963), Murrell (1965), Welford (1971), and Woodson and Conover (1964).

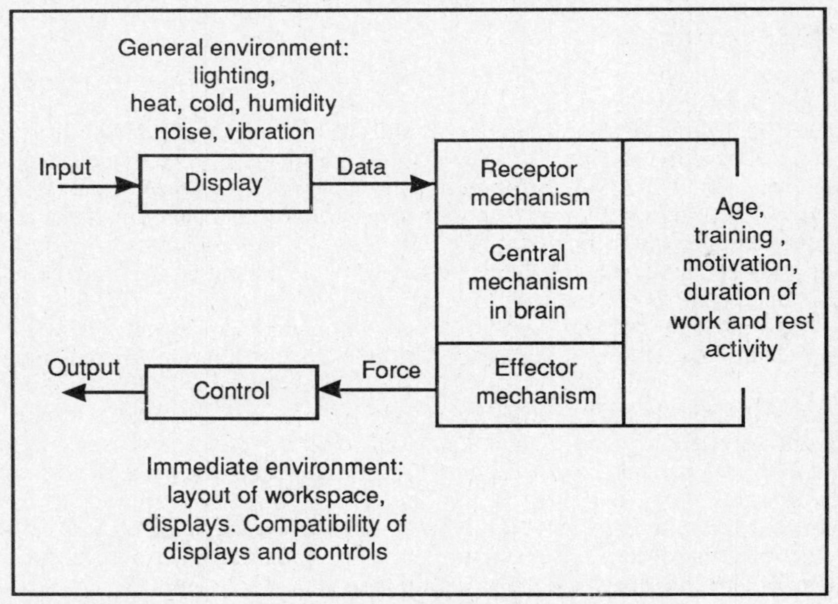

Figure 6.1 A person as an element of a closed-loop system

6.2 People in engineering systems

6.2.1 Occupant of workspace

This is the anthropometric model: data is presented in Section 6.8, based on the charts of Dreyfuss (1967), collected mainly from surveys of the US defence forces. Similar data for Australia does not exist, but we may use the US data with slight corrections where necessary. In Australia, for example, we need to correct for greater girth and hip breadth. In this connection we define two important terms:

Static anthropometry: Body sizes and limb dimensions of the population for whom we are designing.

Dynamic anthropometry: Body and limb movements and space requirements for them. This includes preferred types of control movement, magnitudes of forces and torques which can be applied, arrangements of instruments and control devices such as pedals, levers, handwheels.

An important principle is to design for a range of users, usually from the fifth to 95th percentile, that is, all but the smallest 5 percent and the largest 5 percent. As the data follow a normal curve (*Gaussian distribution*), the fifth percentile point is approximately equal to the mean minus two standard deviations and the 95th percentile is correspondingly the mean plus two standard deviations. Accommodation of 90 percent of potential users is considered a reasonable objective, bearing in mind the trouble and expense required to accommodate the extremes. However, use common sense in applying these limits. An escape hatch or doorway needs to accommodate the 95th percentile man, and a console for a control task needs to have everything within reach of the fifth percentile person.

6.2.2 Source of power

When a person is used to produce power it is usually because the power has to be applied in a variety of different ways at different places (e.g. stacking goods on shelves). The maximum continuous output which a person is capable of depends on which muscles are called into play, but is of the order of 250 watts. A good rule of thumb is that a person should not be required to make a continuous effort greater than about 50 percent of the maximum of which he/she is capable. If the limit is kept down to this level, work can be performed reasonably steadily all day; exertion is kept within the aerobic level — the level at which normal metabolic processes replace the energy consumed without an oxygen deficit.

6.2.3 Sensor or transducer

If we define the function of a transducer as the detection of any change in the physical environment, then we can recognize a corresponding human function. However, in this capacity a person may function in two different ways:

1. *detector* of a signal and initiator of some process in response to that signal;
2. *monitor* of an on-going process where he/she is expected to take some action if a signal indicates a change in the process.

The two tasks are different. In the first case the sensory load tends to be high (in information theory terms), and the function of the human perceptual system is to discriminate and identify the stimuli that are relevant from those that are not. Almost any industrial task serves as an example: assembly work, machining and press operation.

In the watch-keeping task, on the other hand, there is typically a low level of stimulation. The task is to determine when (or if) a signal has occurred. Usually the task is monotonous, for example, inspecting dimensions of industrial components, operating control rooms of power stations or chemical plants and watching radar screens in aircraft control towers.

Errors may occur in both types of task but for diferent reasons. In the first case they are the result of overload, where discriminations between stimuli are too difficult to make or are required too rapidly. In the second case errors may be the result of underloading, when there is insufficient activity to keep the operator alert. The latter situation was described as a 'vigilance' task by the American psychologist, N. H. Mackworth. The classic vigilance effect is a sharp drop in the probability that a signal will be detected after a person has been on watch for half an hour or so.

6.2.4 Processor of information

We have first to quantify the information in incoming stimuli and in human responses to them. To do this we use the special mathematical theory of communication developed by Shannon and Weaver (1949).

Information implies a gain of knowledge in some manner. In order for information to be conveyed there must initially be some uncertainty. The amount of information potentially available increases with the amount of uncertainty in the situation. In this context, information is associated with reduction in uncertainty.

The basic notion of 'information theory' is to measure the amount of information in a series of equally likely outcomes (e.g. successive tosses of a coin) by measuring the total number of binary yes-no decisions that have to be made in order to obtain a precise identification. The unit of measurement is a single binary decision, called a 'bit' (a contraction of binary digit). The bit of information is the amount required to reduce uncertainty by a half. Thus if there are 32 equally likely alternatives, the number of binary decisions required to identify one alternative is five, that is, $\log_2 (32)$.

In general, the amount of information H contained in a set of equally likely alternatives is equal to the logarithm to the base 2 of the number of alternatives n from which the choice is made. Thus,

$$\boxed{H = \log_2 n}$$

The rate at which people can process information depends on how that information is coded (see Miller, 1957). In one experimental task based on readings of ordinary English prose rates of 45 bits per second were observed. If a person is engaged upon a sensorimotor task such as driving a car, then the amount of alphanumeric information he/she can process (e.g. from road signs) is greatly reduced.

It takes people time to react to a signal. Reaction times can vary enormously according to individual differences, level of alertness of the person concerned, level of fatigue, information load experienced and whether or not a signal is expected.

That the correct detailed design of information displays is often crucial to the successful performance of engineering systems is highlighted by an Air Force study of the types of error most frequently made in responding to instruments and signals. Misinterpretation of multipointer instruments (such as some types of altimeter) was the most common error — 18 percent of all errors made. Misinterpretation of direction of indicator movement was next — 17 percent, followed by.failure to respond to warning lights or sounds — 14 percent — and errors associated with poor legibility — 14 percent. The likelihood of such errors being made can be significantly reduced by attention to detailed design.

Murrell (1965) gives a comprehensive set of recommendations regarding the display of information and selection of the appropriate design:

1. *Quantitative reading.* The exact value in some conventional unit has to be available, for example, aircraft altimeter, wattmeter, or digital clock.
2. *Qualitative reading.* An approximate indication of the state of the system is required, for example, high, normal, or low, or within agreed tolerance limits for normal performance.
3. *Check or dichotomous reading.* Is the state of the system O.K. or not? Is the warning light on or off?
4. *Tracking.* A rather special case in which a desired level of performance of a system has to be achieved and maintained by the active control of the operator. It is necessary to provide rate information, for example, a moving pointer on a dial which indicates closure onto the desired level. (See the section following.)

6.2.5 Tracker and controller

The general definition of a tracking task is any situation in which a changing input has to be matched in some way by a system output which is under the control of the human operator. As tracker and controller, a person functions as an element of a dynamic system. The system may be stable or unstable. A stable system tends towards a steady state: if it is perturbed by a change in input it tends afterwards to settle down again, either in a new steady state appropriate to the input it has received or in its former state. If an unstable system is perturbed from its setting by a change in input, it tends to diverge from its previous state instead of settling down. Most systems employing human controllers are neither completely stable nor completely unstable; they are stable within a limited range, and tend towards instability as these limits are approached. An aircraft or motor car is stable within normal operating limits. A ship, a helicopter or a submarine is unstable; without constant attention to the controls, any one of these vehicles will not maintain a steady course.

6.3 Personal factors

6.3.1 General

It is easy to forget that people have attitudes, habits, preferences, emotions, and ambitions. These factors may complicate a design problem but must be considered nonetheless. There is no point in designing a switch or control to be operated in one way when everyone has a strong preference for operating it in another. Some expectations are almost universal and very strong (e.g. the expectation that clockwise movement on a dial indicates an increase). These are known as population stereotypes, and a system which runs counter to them is sure to fail through incorrect operation as soon as the operator becomes flustered or overloaded. Naturally, this would be most likely to occur in an emergency, when correct operation is more important than ever.

6.3.2 Motivation

Why do people behave in the way they do? What are their motives? Attempts by psychologists to come to grips with the complexities of human motivation have given rise to a variety of theories. One of the best known is that put forward by Maslow (1970) who postulated a hierarchy of needs which human beings may strive to satisfy. Once a need toward the lower end of the hierarchy is satisfied, the one above it becomes the most important to an individual.

1. *physiological needs* — food, shelter, sleep, sex;
2. *safety needs* — security, freedom from fears of want, danger, and unemployment;
3. *social needs* — need for love, affection, a sense of belonging;
4. *esteem* — need for self-respect and the respect of others;
5. *need for self-fulfilment, self-actualization.*

Maslow distinguishes between behaviour motivated by deficiency needs, (1) to (4), and that motivated by growth needs, (5). To put the matter briefly, he argues that the satisfaction of growth motivation leads to health whereas the satisfaction of deficiency needs only prevents illness.

Another influential social scientist, Herzberg, has developed a similar theory and applied it to industrial conditions (Herzberg et al., 1959). He argues that there is one set of factors which lead to positive job satisfaction (*'growth needs'*), and another set which, while not involved in positive satisfaction, are sufficient to arouse job dissatisfaction (*'hygiene needs'*). The factors which prevent or end job dissatisfaction are the major environmental aspects of work. Those which produce positive attitudes to work are achievement, recognition, responsibility, and advancement. According to Herzberg's view of their role, managers have a responsibility not only to arrange for the effective use of hygiene factors (e.g. by providing good working conditions) but also to aid the psychological growth of employees and encourage them to become self-actualizers. Cotgrove, Dunham, and Vamplew (1971) describe a case where these ideas were applied to improve the operations of a nylon-spinning plant.

6.3.3 Aesthetics

Whereas Renaissance nudes tend to be bulkier and more lumpy than today's slim pin-ups, Renaissance architecture is still much admired. Our conception of beauty is very subjective and may vary over a period of time. After a long period in the aesthetic wilderness the cast iron lace balconies on Victorian houses around Melbourne and Sydney are now recognized as being of distinctive quality.

It is difficult for the engineer to find a firm foundation in the shifting sands of fashion and taste. A lack of response to aesthetics sometimes induces hostile attitudes in community leaders. In one controversy in England regarding the construction of electricity transmission lines across a stretch of beautiful countryside, the editor of a leading newspaper, *The Times*, addressed the following question to the head of the electricity authority: *'When electricity pylons are being evolved, are designers concerned with aesthetics brought in or is this work left to the engineers?'* The implications of this question are obvious.

One of the prime uses of aesthetics in the design of an engineering product is to indicate function and purpose. Simple, easy-to-perceive forms which recognize and reflect function and purpose automatically do this. A suburban train travels forwards; it should look as though it does in fact do this. Long horizontal lines should be emphasized in the appearance of its carriages instead of arrays of short vertical lines corresponding to the openings for doors and windows.

Important concepts relevant to discussions on aesthetics are :

1. *form and proportion*;
2. *visual balance* — achieved by a subconscious association with the principles of physical balance, for example, weights and levers;
3. *clarity of form and expression* — lack of clutter, a product looking as though it will do what it does;
4. *colour*;
5. *rhythm* — an effect gained by repetition, for example, a balustrade;
6. *surface texture*;
7. *unity* — a coherence of parts forming the whole, achieved through some fundamental similarity.

For further reading see Ashford (1969). An interesting study of the adaptation of form to purpose in living creatures and plants is that by Thompson (1961).

6.4 Environmental factors

Important environmental factors are: air temperature and humidity, air velocity, illumination, noise and vibration. Any discussion of these matters must be based on a firm foundation of specialized branches of engineering science, which is outside the scope of this book. Murrell's treatment (1965) is particularly helpful.

6.5 Safety

6.5.1 Theory

We describe an accident as an occasion in which an engineering system comprising people and hardware moves outside its limits of stable operation. Figure 6.2 represents the time history of an accident.

If we examine the history of an accident we find there is a chain of events; the chain of events invariably involves human factors, physical factors associated with the equipment being operated, and environmental factors. To take a hypothetical example, imagine what happens when four people are crowded into the front of a utility truck, driving at night, it's raining and the windscreen wiper doesn't work too well, the tyres are bald, and a pedestrian in a dark suit steps out from the pavement. What is the cause of the ensuing accident? In fact there are many causes.

Because there is no single cause of an accident there is no single solution to the problem of preventing that accident happening again. We have to look for ways of attacking the chain and preferably of making the system 'fail safe' — in other words, a tendency for the system not to move to the right (Figure 6.2) or, if pushed to the right, to come back again. We have to look at the people involved, the equipment they are using and the environment they are in; then identify the critical factors and try to improve them, so that in an emergency there is a greater probability of the system remaining stable.

Stable operation	*Metastable operation*	*Unstable operation*	*Energy exchange*	*Repair or scrap damaged parts*
System under control	System disturbed, but control may be regained	Control cannot be regained	Impact. Damage to people and property	
TIME →				

Figure 6.2 Time history of an accident

To talk glibly of 'human error' is to take the easy way out (an example of mental laziness). What we have to ask is: Why did the person concerned act in a particular way in the chain of events leading up to the accident? Was the person being asked to do something for which inappropriate or insufficient information was provided, or for which the training was inadequate? Was the person put in a position in which the system in normal operation was so near to being unstable that it just shot to the right at the first disturbance?

6.5.2 The cost of safe design

The engineer is responsible for committing the community's resources to the design and construction of safe devices and systems. This responsibility implies an awareness of the size of the commitment and of the results to be achieved in proportion to the resources committed.

Even very rough indexes of comparison can reveal orders of magnitude differences, for example, between surface and air transport. One such index would be an estimate of the dollars spent to ensure a safe design in proportion to the number of lives saved. One might ask what the relative values of the following are:

1. crash helmets for motor cyclists;
2. seat belts in automobiles;
3. inflatable rafts in commercial aircraft flying over oceans;
4. firefighting equipment at airports.

6.6 Design for handicapped people

Little systematic research has been carried out into the capacities and requirements of people handicapped by physical disability or disease. The one precept that can be laid down with certainty is that the design of equipment to be used by handicapped people should be undertaken *with* the people concerned and not *for* them.

Goldsmith (1967) has considered in great detail '*architectural design criteria for people with physical disabilities who are handicapped by buildings*'. Although mainly concerned with people in wheelchairs, his book is an essential starting point for any designer in this field.

6.7 Notes from a designer's workbook

6.7.1 An example of ergonomics in engineering design

The design of a student work area

A heritage of the Industrial Revolution is the separation of people from their working environments, witness Henry Ford's assembly lines. Instead of a union of thought, word, and deed, there is a gap which all too often leaves the human spirit unfulfilled. If we stand back and look at the factories and offices of contemporary industrial societies, can we be satisfied that they acknowledge people's emotional and spiritual needs? *'Is life no more than this*?' cried a distraught King Lear. Any human enterprise should involve the full person, but too often we accept the tawdry and second-rate, perhaps because of a lack of resources to do more but more likely because of a failure of creative imagination. But these are emotive words and we should remember that, over time, human aspirations change and grow; one generation's solutions to its problems become the source of the next generation's problems.

This philosophical introduction to an example of ergonomics in engineering design is prompted by Figure 6.3: a photograph of a student work area designed and constructed in the late 1940s to provide urgently-needed accommodation in the University of Melbourne for the bulge in enrolments after World War II. To today's eyes this area has an angular, monotonous, uninspiring appearence. Be that as it may, with the passage of time, the 1940s and 1950s accommodation crisis passed, and the next generation recognized a need for refurbishment and renewal. Since the area was to be used by engineering students for personal study, design assignments, report writing and associated activities, a student design competition was held.

Figure 6.3 Student work area at Melbourne (*circa* 1950)

Sketches and diagrams from one leading entry are exhibited in Figures 6.4, 6.5, 6.6 and 6.7.

Figures 6.4 and 6.5 show how careful attention to detailed anthropometric data leads to dimensions of the individual workspace, which in turn comprises a desk with drawers, cupboards, adjustable board, and personal lighting. There are vertical screens behind each desk to provide space for pinning notices, assignment sheets, sketches, and drawings. These vertical partitions are a contrasting element of the visual landscape, and at the same time help to give each student a sense of privacy, even of territorial ownership. Figure 6.6 indicates the generally pleasing appearance of entrance areas and aisle, another contribution to the aesthetic pleasure provided by the proposed design.

To sum up, the successful synthesis of ergonomics, aesthetics and technology present major challenges to the ingenuity and creative talents of engineering designers. This example exhibits some of the thinking necessary to meet these challenges.

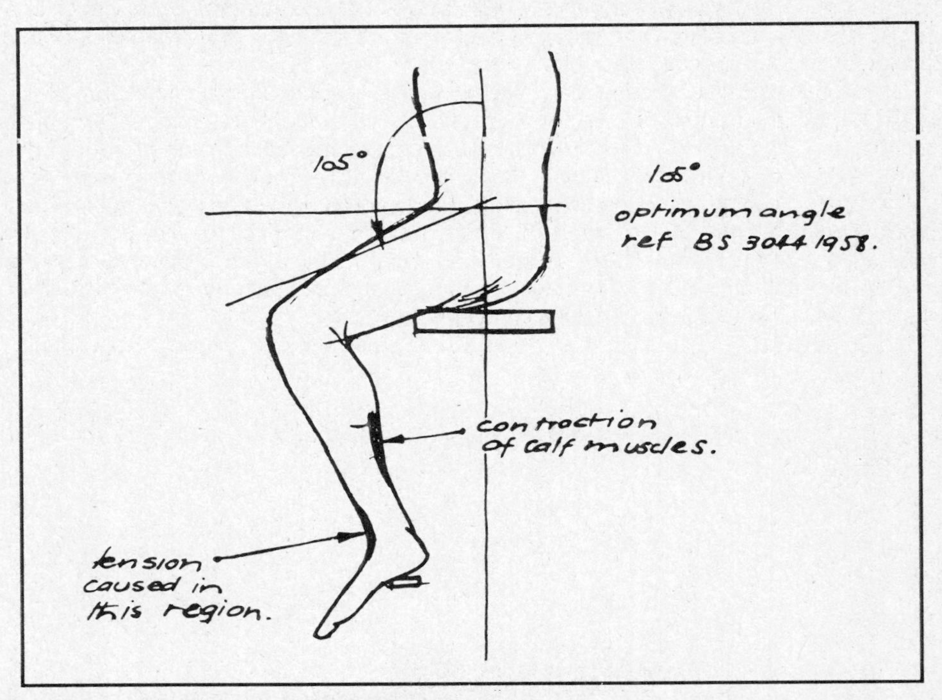

Figure 6.4 Sketches of seated operator-elevation

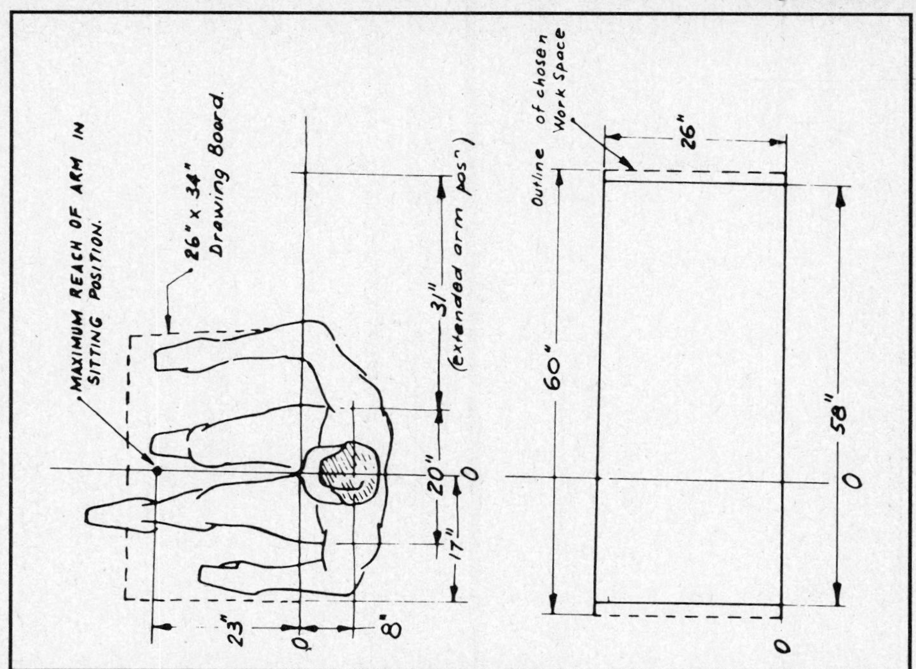

Figure 6.5 Further layout sketches for workspace

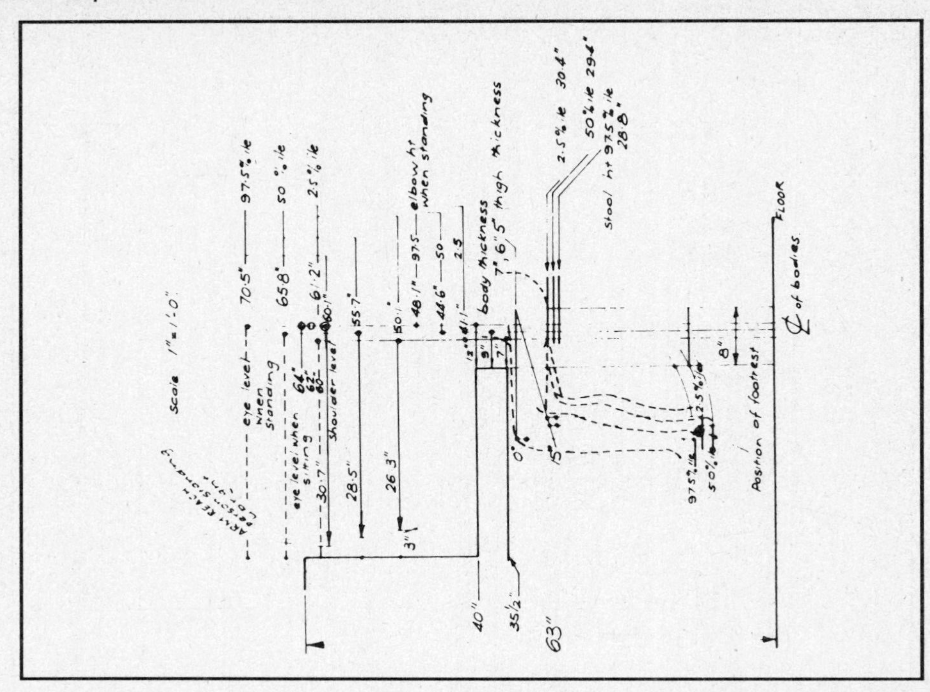

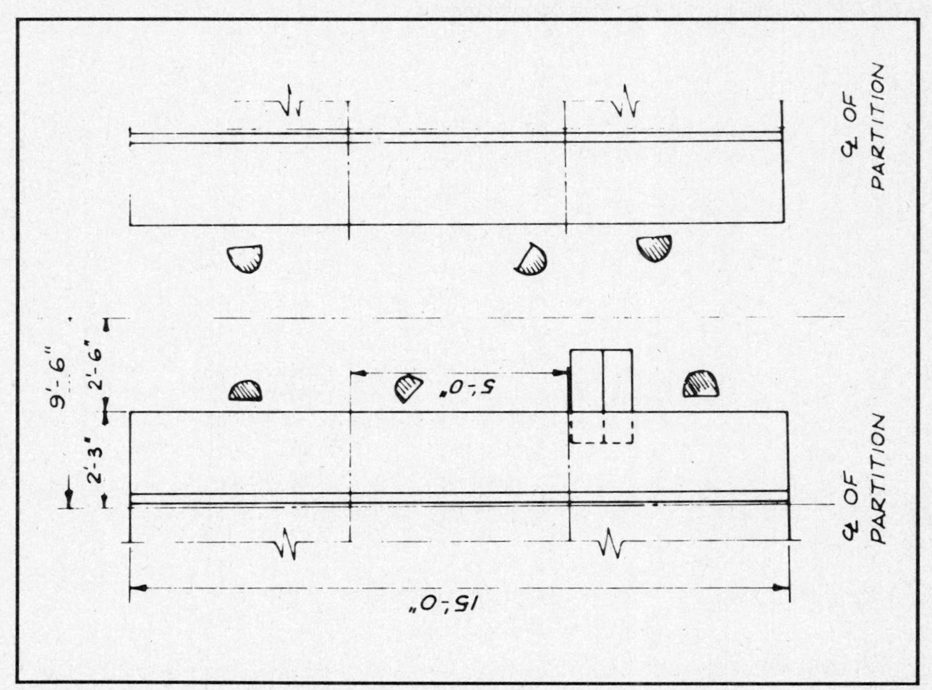

Figure 6.6 Further layout sketches for workspace

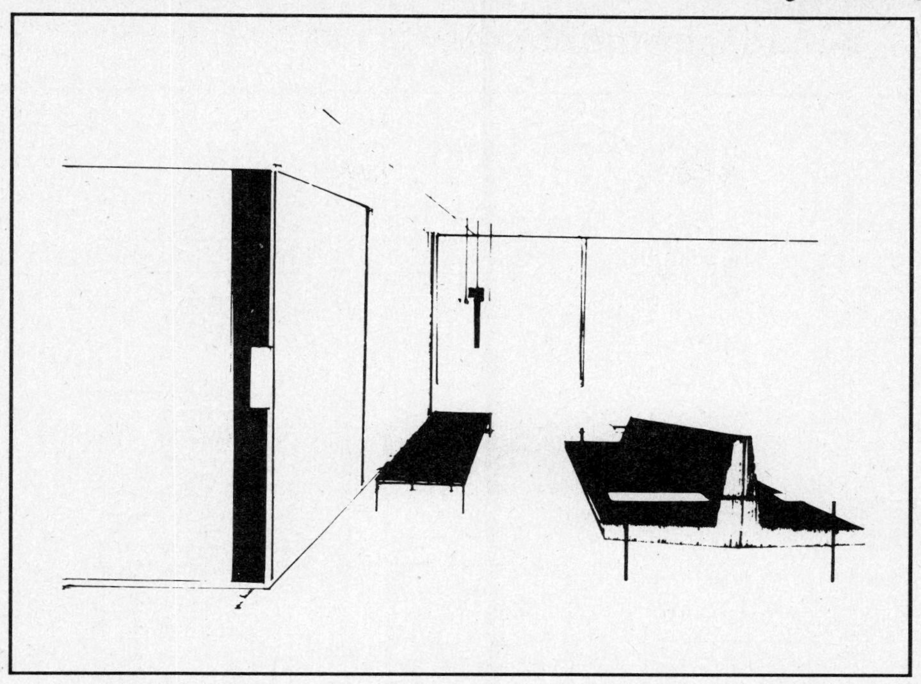

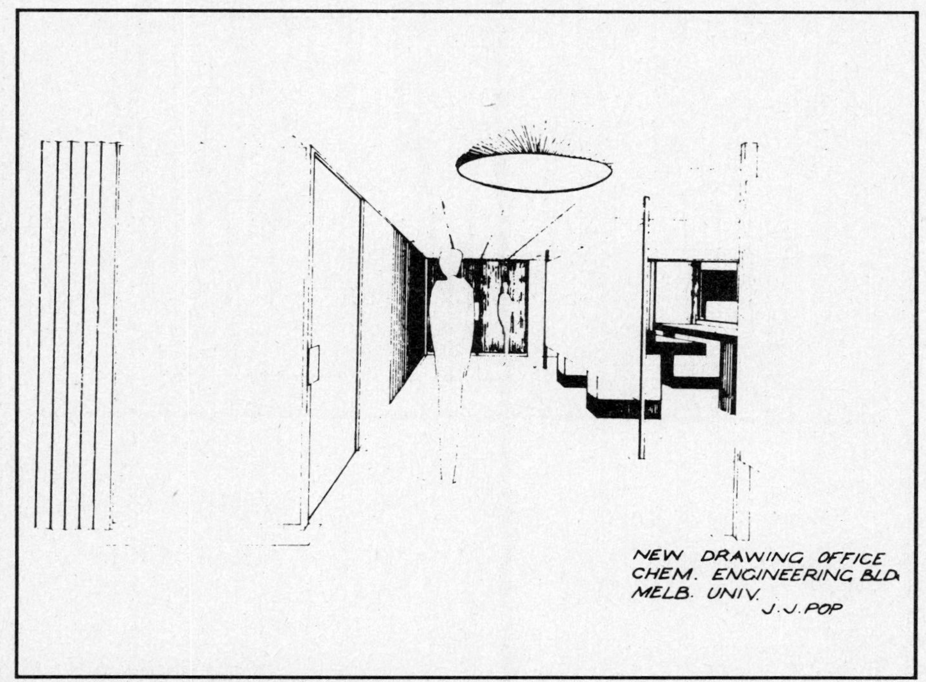

Figure 6.7 Entrances and aisleways

6.8 Anthropometric data (Dreyfuss, 1967)

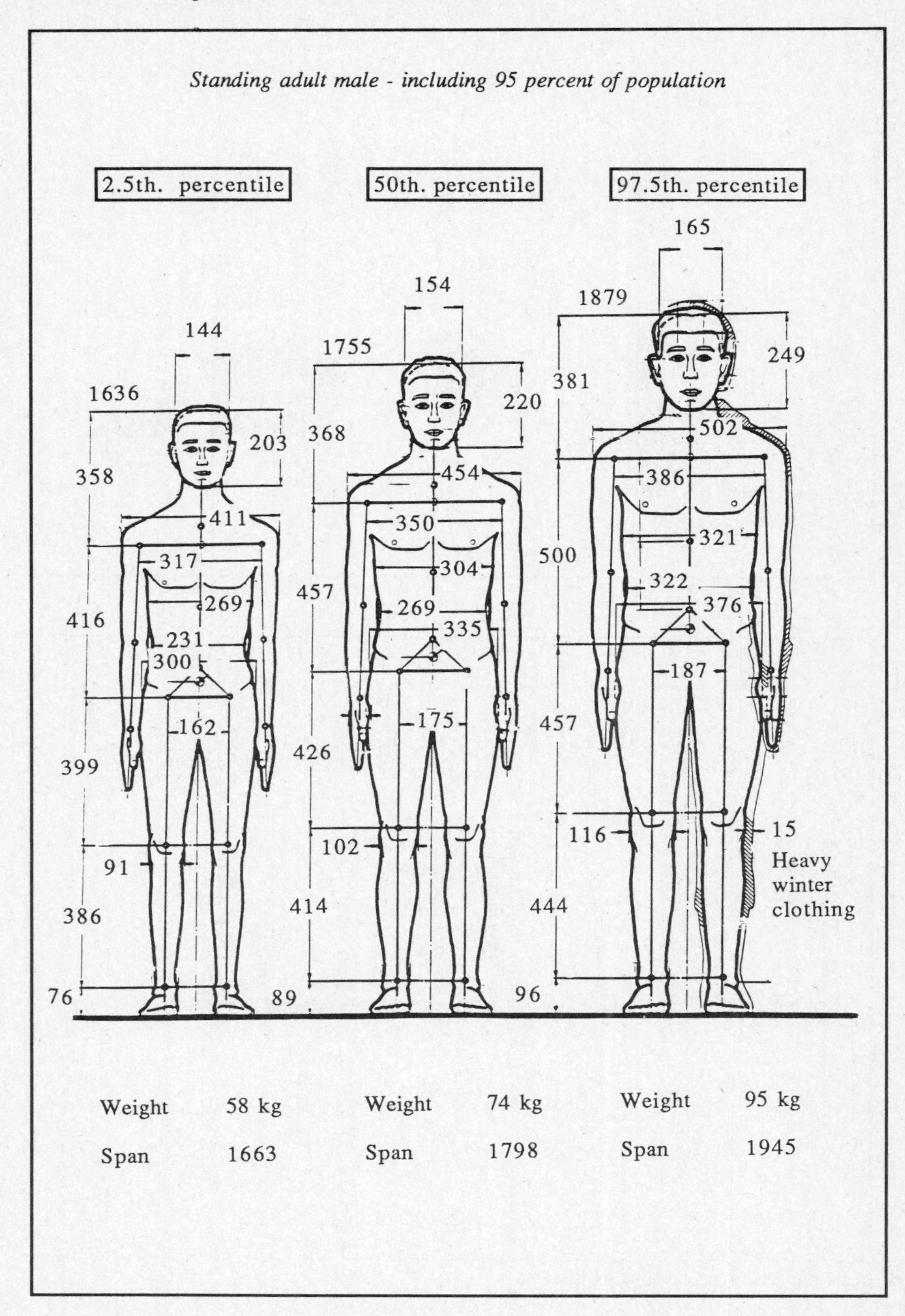

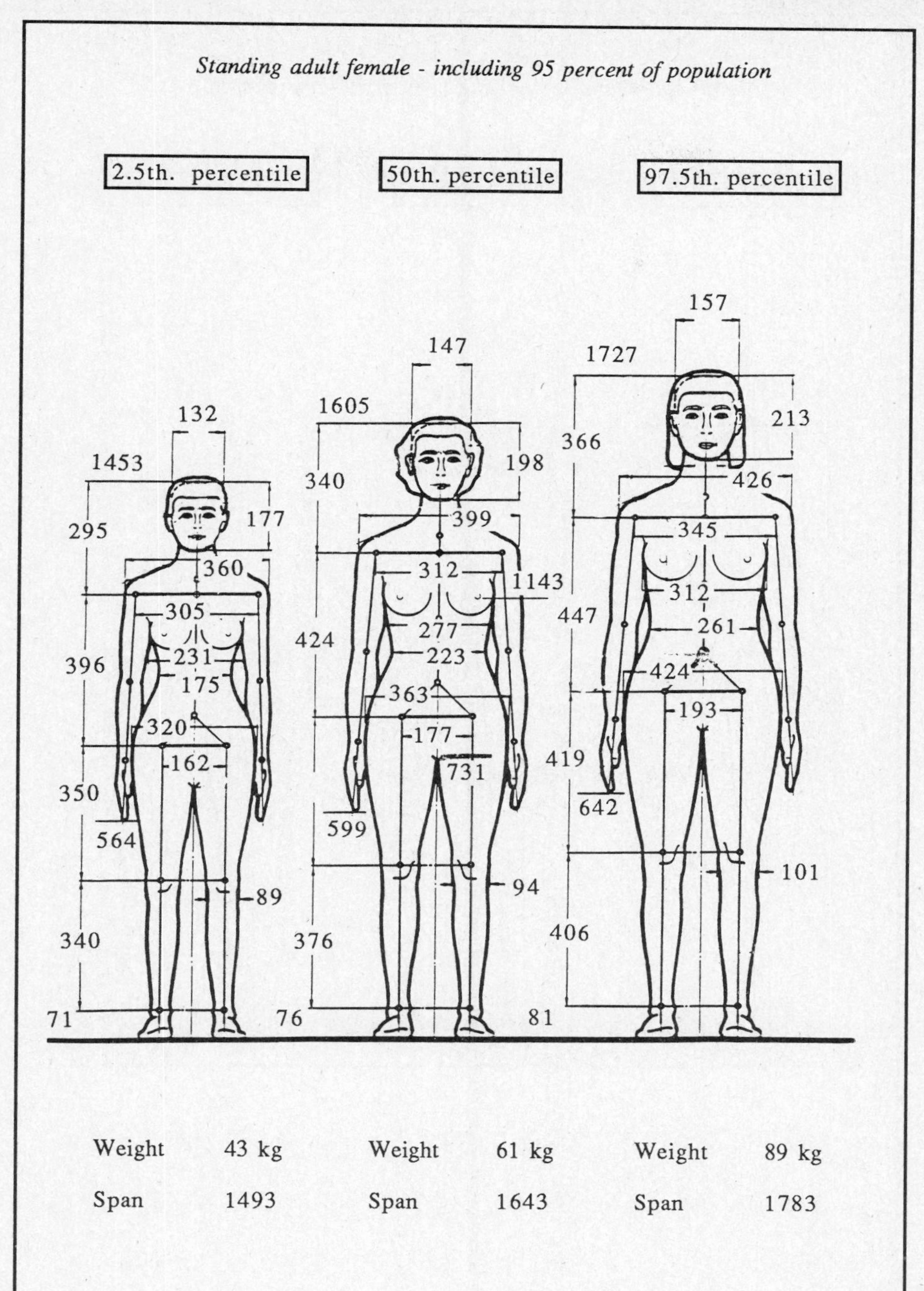

Weight 43 kg
Span 1493

Weight 61 kg
Span 1643

Weight 89 kg
Span 1783

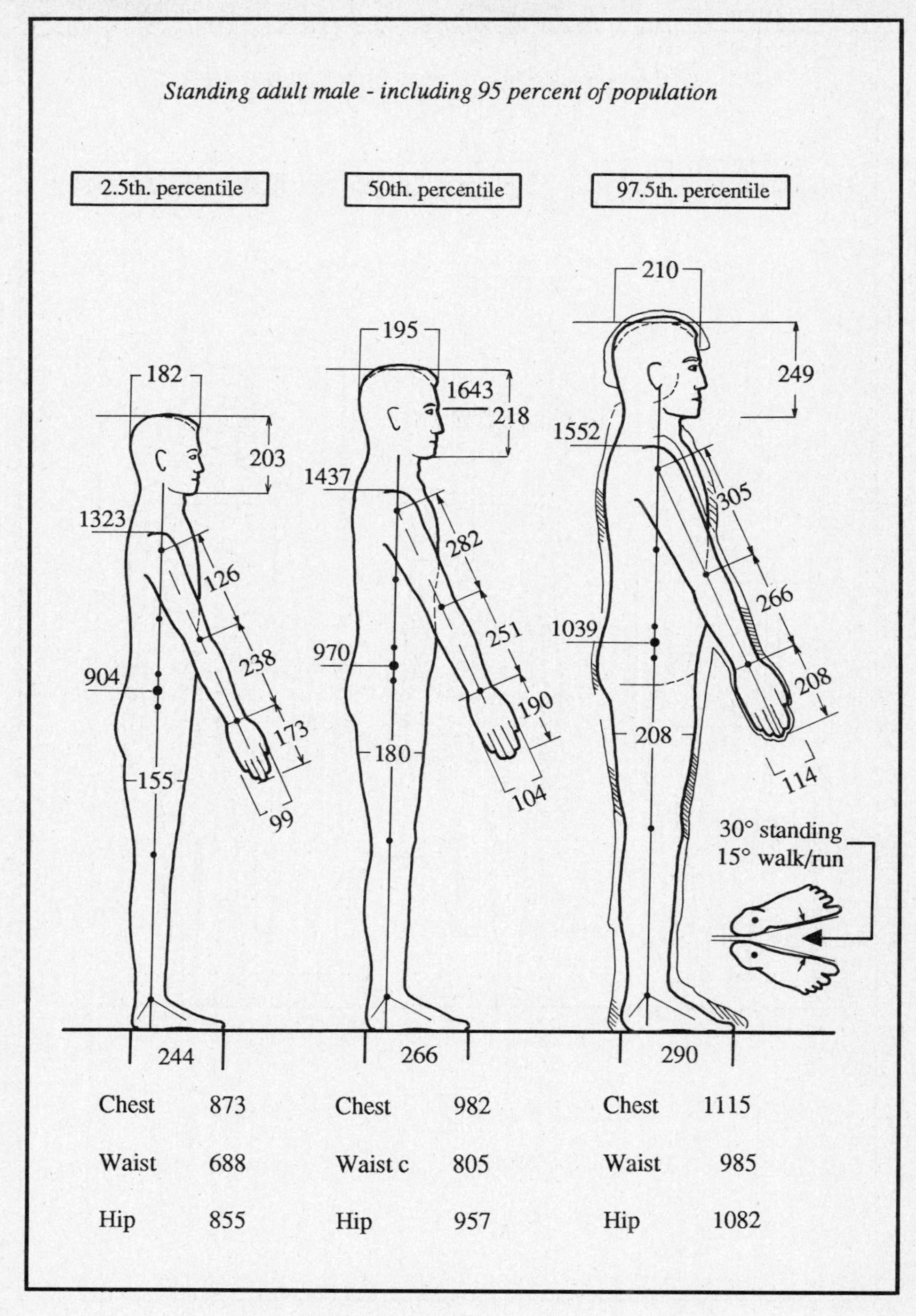
Standing adult male - including 95 percent of population
2.5th. percentile
50th. percentile
97.5th. percentile
182
203
1323
126
238
904
173
155
99
244
195
1643
218
1437
282
251
970
190
180
104
266
210
249
1552
305
266
1039
208
208
114
30° standing
15° walk/run
290
Chest 873
Waist 688
Hip 855
Chest 982
Waist c 805
Hip 957
Chest 1115
Waist 985
Hip 1082

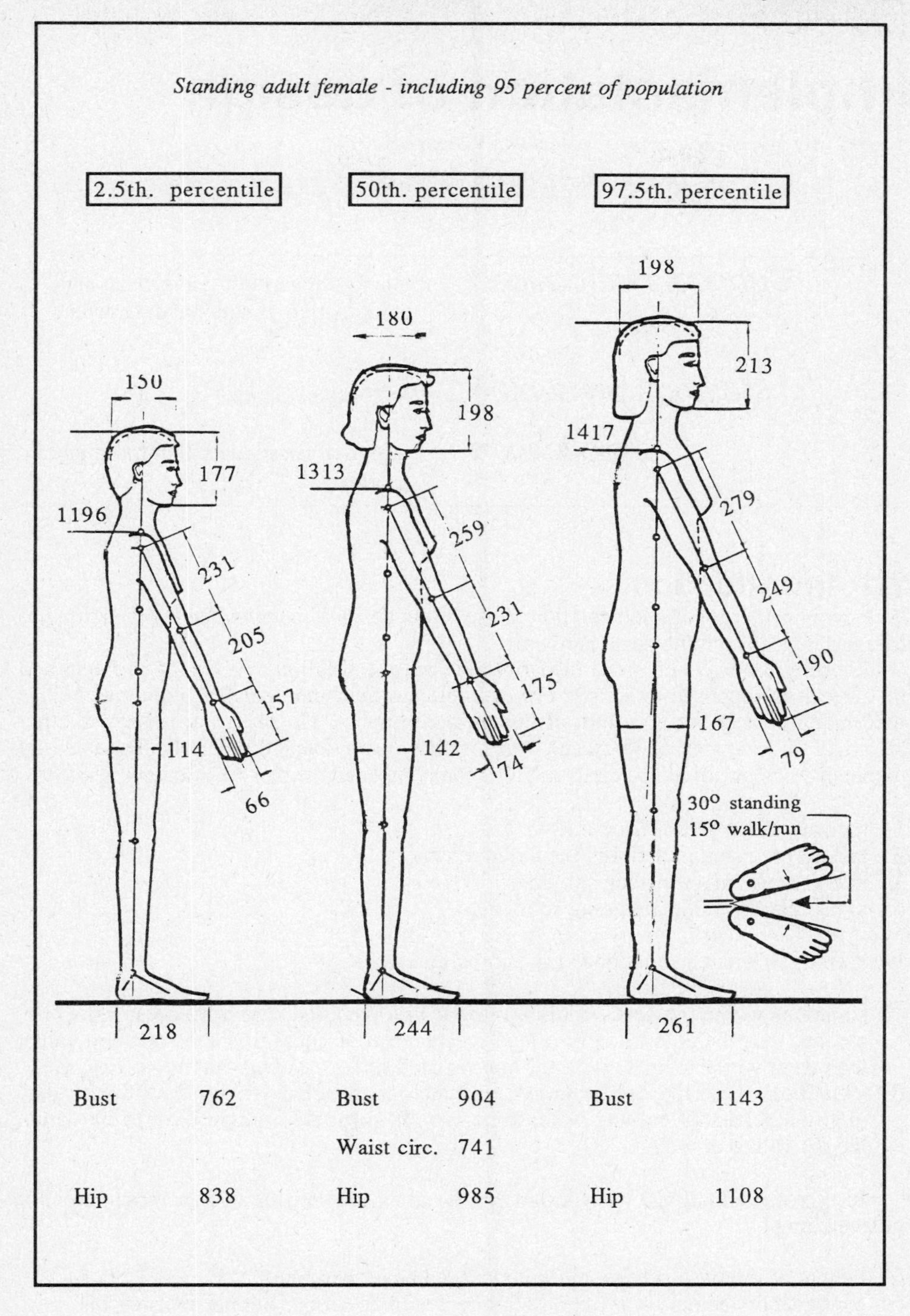
Standing adult female - including 95 percent of population
2.5th. percentile
50th. percentile
97.5th. percentile
150
177
1196
231
205
157
114
66
218
180
198
1313
259
231
175
142
74
244
198
213
1417
279
249
190
167
79
261
30° standing
15° walk/run
Bust 762
Hip 838
Bust 904
Waist circ. 741
Hip 985
Bust 1143
Hip 1108

Chapter 7

Implementation of design

One way to recognize error is the fact that it's universal.

Jean Giraudoux, French Playwright

Concepts introduced	mismatch, variability in design, and manufacture, functional dimension.
Methods presented	loop equations, risk analysis.
Application	design of passive oscillator, design of gearbox.

7.1 Introduction

We now consider the implementation phase of the design process when the designer has to translate ideas into physical hardware.

There will always be some disparity between the solution to a design problem and the need it is supposed to satisfy. For example, no machine operated by human beings will match exactly the spectrum of human performance. No source of power, whether electrical, mechanical, or whatever, will match a given demand precisely. The inability to match a design solution exactly to fulfil some intended need or requirement is due to:

1. the variability of the requirement;
2. lack of precision in defining the requirement;
3. uncertainties in specifying solutions;
4. uncertainties in implementing solutions.

We classify mismatch situations in two important ways:

(a) situations where the need–solution mismatch is specified at one boundary only (e.g. a beam must be able to carry a load greater than or equal to some specified value; the rate at which a fan delivers air must exceed some specified minimum flow rate).
(b) situations where the need-solution mismatch is specified within some finite range (e.g. the acidity of cooling water used in a chemical plant might have to lie within the pH values of 6.5 and 7.0).

In type (a) situations, over-design can be accepted and the designer can cater for mismatching by:

(i) factors of safety — where any failure would be catastrophic;
(ii) contingency factors — where any failure would be serious but not catastrophic.

The experienced designer can associate either formally or informally (intuitively) a probability distribution with each of the design variables. As the designer's experience and practice in a particular type of situation increase, so will any factors of safety be closer to unity or contingency factors closer to zero. However, in a new design for which there are few or no precedents, there will be some level of risk attached to the designer's estimates of the variables and possibly also to the associated probability distributions. Larger safety factors may then be used, or special risk analyses carried out (see Section 7.3).

In type (b) situations, it is usually easier to inspect the system or component to check for a mismatch condition than it is in the type (a) case. Consequently, action is taken to ensure that any mismatch detected during the manufacture of the system or component which is outside the limits or tolerances specified leads to the rejection of that system or component.

From these general considerations we now turn to consider the detailed design of engineering components. One important class of type (b) situations is concerned with tolerancing the dimensions of such components.

7.2 Tolerances in the detailed design of components

In the design of engineering components, the designer must be aware of the cumulative effects of errors in the dimensions and shapes which comprise the component geometry. These errors may result from imprecise manufacture or imprecise measurement of dimensions.

The property of the component for which the limits of error (mismatch) are prescribed is known as the functional requirement. If this requirement is a geometric feature then it is known as the functional dimension. Figure 7.1 illustrates this.

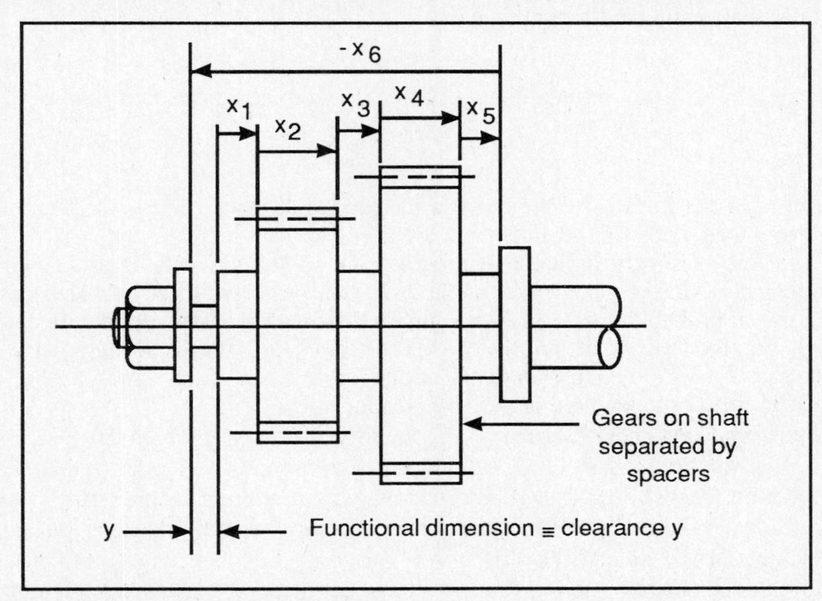

Figure 7.1 Example of a functional dimension

In the usual notation, functional dimensions are designated y_r, with clearances positive ($+ y_r$) and interferences negative ($-y_r$), see Fortini (1967). The detailed dimensions are designated x_r, and in Figure 7.1 are measured positive if towards the right.

The fundamental properties of the x_r are:

1. they are all parallel;
2. each x_r is a dimension between two parallel surfaces on the same detail;
3. the surfaces at which each x_r begins and ends are surfaces that mate with adjacent details;
4. there is only one x_r for each detail;
5. together with y_r the x_r form a closed loop known as the dimension path or loop.

$$\sum (x_r + y_r) = 0$$

Dimension loops are used to determine the effect of detail errors on functional dimensions. Thus, suppose the error in a typical dimension is t_r so that the dimension lies between x_r and $x_r + t_r$ (unilateral tolerance). The question then arises as to what are the limits of error (mismatch) on the functional dimension. There are several approaches to this problem, the most common being 'worst case' analysis. The pessimistic assumption is made that all component dimensions are at their worst limit. In Figure 7.1, one worst limit (the largest clearance) would correspond to x_1, x_2, x_3, x_4, x_5, being on the low limit and x_6 being on the high limit. The highest value of the functional dimension (clearance y) is then:

$$y_h = x_6 + t_6 - x_1 - x_2 - x_3 - x_4 - x_5 \quad \textbf{(7.1)}$$

By similar reasoning, the lowest value is:

$$y_l = x_6 - x_1 - x_2 - x_3 - x_4 - x_5 - (t_1 + t_2 + t_3 + t_4 + t_5) \quad \textbf{(7.2)}$$

If the variation in y is t_y, subtracting equation (7.2) from equation (7.1) yields:

$$t_y = y_h - y_l = \sum t_r \quad \textbf{(7.3)}$$

Thus, in the worst limit case, the error in the functional dimension is the sum of all the worst errors that appear in the dimension loop.

In practice it is very unlikely that each detail dimension will be at its worst limit simultaneously. More sophisticated methods of analysis have therefore been developed making use of probability theory. A probability distribution is associated with each detail dimension on the basis of the known characteristics of the manufacturing process used to produce that dimension. Gaussian distributions (or versions of them such as the 'moving normal') are often encountered in such applications.

Engineering designers frequently encounter the inverse problem to that discussed above, namely, 'What values of t_r should we assign to the x_r, given a specified value of t_y?' Equation (7.3) still holds and there is in theory an infinite number of t_r which will satisfy it. Ideally the designer should allocate t_r in a way which gives rise to least cost of manufacture. To be able to do this, the relationship between any t_r and the cost of achieving that t_r must be known, that is:

$$\text{Cost} = C_r(t_r) \quad \textbf{(7.4)}$$

A number of authors and research workers have suggested that equation (7.4) should be an inverse power relationship, thus:

$$C_r = k\, t_r^{-n} \tag{7.5}$$

However, the designer has frequently to rely on judgment, experience, and knowledge of manufacturing processes.

7.3 Risk analysis: Uncertainties in design variables

We saw in the previous section how worst case analysis could be used as a simple means of circumventing the problem of dealing with uncertainties in the values taken by design variables. Sometimes the designer has formally to take account of these uncertainties, and the following discussion of risk analysis is intended to show how this may be done.

Risk analysis, then, is a method for handling uncertainties. For example, in deciding whether to implement a new project, the criterion C we adopt — net present value or net rate of return on investment — will be affected by a number of factors (x_i) which we can 'quantify but whose exact values we may not be completely certain about.'

Suppose that:

$$C = f(x_1, x_2, x_3, \ldots . x_i, \ldots . x_n)$$

where each x_i has an associated probability density function, as in Figure 7.2 for example. The probability that the value of x will lie between the given limits is given by:

$$Pr(0<x<X) = \int_0^X f(x)\,dx$$

If we take the most likely value of each variable (X' in Figure 7.2) and compute the criterion, we are restricting our judgment to one numerical result. A complete judgment should take into account the possible range of each variable and the likelihood of each value within this range. This implies that we should associate a probability distribution with each variable; if the project has few or no precedents the distributions would be

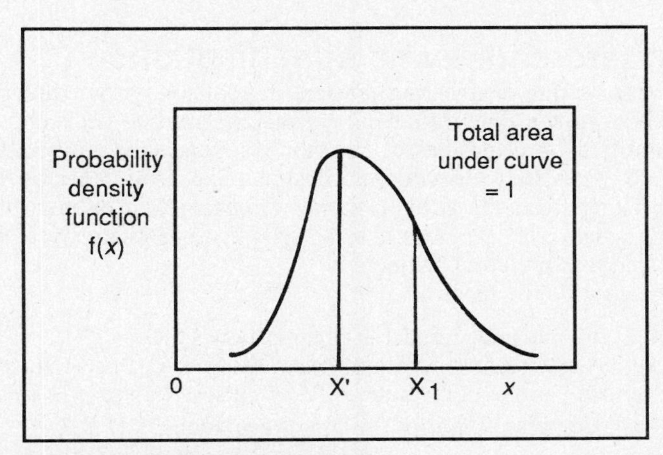

Figure 7.2 A typical probability density function

based on the considered opinions of experts and would thus be subjective. The subjective probability distributions could then be manipulated according to classical statistical theory. In any case the aggregation of the individual probability distributions for the x_i's to find the probability distribution for C is carried out by means of a Monte Carlo simulation (Krick, 1969, p. 61).

Pouliquen (1970) gives an example of the application of this method to the assessment of a project consisting of the construction of a new breakwater and harbour facility. The criterion chosen was the net rate of return on investment, which was found to depend on seven variables:

1. capital cost of the project;
2. productivity of labour (tons of cargo per gang hour);
3. average value of a ton of cargo;
4. percentage of cargo tonnage saved through reduction in damage;
5. rate of growth of imports;
6. value of a ship working day;
7. life of the assets.

After consultation with experts, a probability distribution was assigned to each of these variables and the cumulative probability distribution of the rate of return *r* computed. The results were as follows:

Mean value of *r*	= 10.6%			
Standard deviation of *r*	= 2.5%			
Probability that *r* would exceed:		5.0%	=	0.99
		7.0%	=	0.94
		8.0%	=	0.85
		10.0%	=	0.61
		15.0%	=	0.02

As the chance of getting a rate of return of less than 8 percent was estimated as 15 percent and as there was a better than even chance of getting more than 10 percent, it was decided to go ahead with the project.

7.4 Standard codes and specifications

Standards set out current engineering practice in common design situations. In effect, they save the designer's time and effort by making routine decisions. Unnecessary variety and proliferation of designs of common components is prevented. However, in more advanced, sophisticated work the designer has to be careful that excessive devotion to existing standards is not a barrier to engineering improvements. Where a range of performance characteristics is to be specified quantitatively, it will in general be based on a series of preferred numbers.

Standards authorities include:

1. international (International Standards Organisation: I.S.O.);
2. national (e.g. Standards Association of Australia: S.A.A.; British Standards Institute: B.S.I.; American Petroleum Institute: A.P.I.; Deutsche Industrie Normen: D.I.N);
3. domestic (e.g. Victorian[1] Uniform Building Regulations: V.U.B.R.);

---

[1] Refers to a State authority standard in Victoria, Australia.

4. individual engineering organization, company, or government agency.

There are five important types of standard, namely:

1. *Dimensional* standards to secure interchangeability, as in a nut or bolt, and to eliminate unnecessary variety of types for the same or similar purposes.
2. *Performance* or quality standards to ensure that a finished article or product is fit for the job it has to do (e.g. there are performance standards for crash helmets and car seat belts). Materials standards are also included under this heading: they specify the chemical composition, heat treatment, physical and mechanical properties of different materials.
3. Standard *methods of test* enable materials or products to be compared on a uniform basis. Such test methods assess, for example, strength and hardness of metals, elasticity of rubber, viscosity of liquids, and similar key characteristics of materials used in engineering. Test methods have often to be developed before a performance standard can be prepared, for example, for the rating of internal combustion engines.
4. Standard *technical terms* and *symbols* provide a common, easily understood language for engineering industry and commerce.
5. Standard *codes of practice* set out methods for the design, installation, and operation of equipment such as pressure vessels, lifts, and power transformers, to name a few examples.

7.5 Interaction of design with manufacturing and construction

The designer has to interface successfully with the people responsible for manufacture and construction. A brief review of manufacturing processes and their influence on design now follows.

Major manufacturing processes are set out below:

1. *casting* — forming from the liquid;
2. *mechanical working* — forming from the solid:
   (a) hot working (e.g. hot rolling, forging, extrusion);
   (b) cold working (e.g. cold rolling, cold extrusion, pressing);
   (c) forming from powder (e.g. sintering, plastic moulding);
3. *machining* — cutting from the solid (e.g. turning, drilling, boring, shaping, slotting, milling, grinding, broaching, sawing, reaming);
4. *fabrication* — joining parts together (e.g. welding, brazing, soldering, adhesives);
5. *heating treatment* (e.g. annealing, normalizing, quenching, tempering);
6. *surface treatment* (e.g. shot blasting, painting, plating, etching);
7. *assembly* (e.g. manual assembly, robotics);
8. *quality assurance* (e.g. inspection and testing).

The following factors have to be borne in mind when deciding what manufacturing process should be used in a given application:

1. selection of material;
2. size of part;
3. rate of production;
4. investment in manufacturing plant;

5. physical properties of the parts produced by the process (e.g. lightness, strength);
6. complexity of form;
7. section thickness;
8. dimensional accuracy;
9. subsequent processes;
10. appearance and surface finish;
11. costs of raw materials, defective components, scrap rate.

7.6 Notes from a designer's workbook

7.6.1 Design for variability in component performance — electronic circuit

Manufacturing errors can and do occur in all industries. The designer must be aware of the levels of such errors to account for them during the design process. The following example and that in the next section, drawn from two very different environments, are intended to illustrate typical ways in which designers deal with errors arising from variability in manufacture which affect component performance and geometry.

Figure 7.3 shows a circuit for a full wave rectifier with smoothing filter which is to be used as a direct current power supply and manufactured in large quantities. Analysis of the circuit yields the following two relationships:

1. Secondary circuit:

$$V_{DC} = (V_s - 2V_d) \times \left\{ 1 - \frac{1}{2f R_L C_F} \right\} \tag{7.6}$$

2. transformer characteristics:

$$V_{IN} = 24V_s \tag{7.7}$$

C_F = filter capacitance (farads).
f = frequency of a.c. voltage supply (hertz).
R_L = load resistance (ohms).
V_d = voltage drop across one diode while conducting (volts).
V_{DC} = output direct current voltage (volts).
V_{IN} = mains supply voltage (a.c. volts).

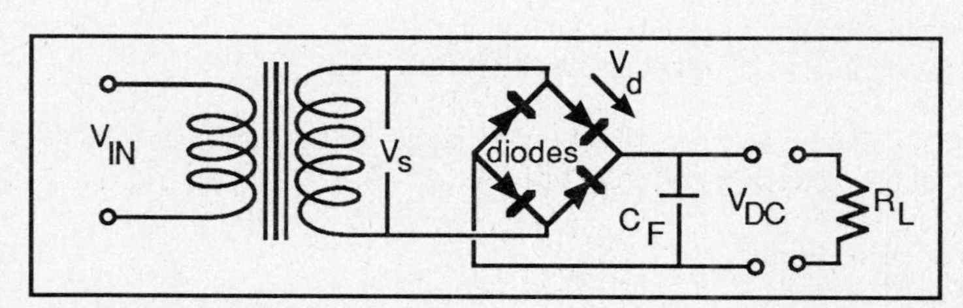

Figure 7.3 Transformer full wave rectifier circuit

The average output voltage (V_{DC}) must be regulated to 8.0 ± 0.5 volts with the following input conditions:

$$V_{IN} = 230 \text{ to } 240 \text{ volts}$$
$$f = 50 \text{ or } 60 \text{ hertz}$$

In addition, diodes with the characteristics V_d = 0.5 to 0.7 volts will be used. Although the nominal value of R_L is 10,000 ohms, it is desirable (for reasons of versatility) to design a circuit permitting a fairly wide range of loads. Capacitors of any nominal value can be produced with ± 10 percent, ± 5 percent or ± 1 percent tolerances. The 1 percent tolerance components are very expensive, but the 5 percent capacitors are only slightly more expensive than the 10 percent ones.

Problem

1. Find values of C_F and its tolerance to satisfy the design requirements.
2. Find the range of R_L which can be used with the supply.

Solution

Substitute equation (7.7) into equation (7.1):

$$V_{DC} = \left\{ \frac{V_{IN}}{24} - 2V_d \right\} \times \left\{ 1 - \frac{1}{2fR_LC_F} \right\}$$

Let: maximum value of $C_F = C$
minimum value of $C_F = C - c$
maximum value of $R_L = R$
minimum value of $C_F = R - r$

1. For maximum value of V_{DC} the following condition is true:
V_{IN}, f, R_L, C_F are at their maximum; V_d is at its minimum;
hence:

$$8.5 = \left\{ \frac{240}{24} - 2 \times 0.5 \right\} \times \left\{ 1 - \frac{1}{2 \times 60\ RC} \right\}$$

$$RC = 0.150 \qquad \textbf{(7.8)}$$

2. For minimum value of V_{DC} the following holds:
V_{IN}, f, R_L, C_F are at their minimum; V_d is at its maximum;
hence:

$$7.5 = \left\{ \frac{230}{24} - 2\times 0.7 \right\} \times \left\{ 1 - \frac{1}{2\times 50(R-r)(C-c)} \right\}$$

yielding: $(R-r)(C-c) = 0.120$ **(7.9)**
The cheapest capacitor has: $c = 0.2C$;
substitute into equation 7.4 $(R-r)\ (0.8C) = 0.120$ **(7.10)**

Divide equation (7.10) by equation (7.8): $\frac{R-r}{R} = 1.0$

therefore: $r = 0$

Since $r = 0$ is not practicable, the next lowest tolerance capacitor must be used.

Hence, if $c = 0.1C$: $(R-r)(0.9C) = 0.120$

therefore $\frac{R-r}{R} = 0.89$

$r = 0.11R$

Average load resistance is: 10,500 Ω

$R = 10{,}500\ \Omega$

Substitute into equation (7.3): $C = \frac{0.150}{10500} = 1.43\text{x}10^{-5}$ Farad

Specification

Nominal capacitance = $1.36 \text{ x } 10^{-5}$ F(±5%)

Load resistance range: 10,500 Ω to 9,500 Ω

(Range with 1% capacitors is approx. 11,000 Ω – 9,000 Ω)

7.6.2 Design for variability in component geometry — gearbox

Problem

1. Figure 7.4 is a full section of a gear box. What are the dimensions and tolerances required on the length of the gear wheel's boss to maintain the specified axial clearance between gear wheel and bushes of 0.05 to 0.40 mm?
2. Make a neat isometric sketch of the main gearbox casing when viewed from the upper right-hand corner of Figure 7.4.

Solution

Referring to Figure 7.4, the following equations may be written:

Path equation: C = B – A – E – 2D

Minimum C = Min. (B) – max. (A + E + 2D);

Maximum E = Min.(B– C) – max. (A + 2D) = 57.50 – 0.05 – 20.00 –6.5

= 30.95

Maximum C = Max.B – min.(A + E + 2D)

Minimum E = Max.(B– C)–min.(A + 2D) = 57.65 – 0.40 – 19.95 – 6.40

= 30.90

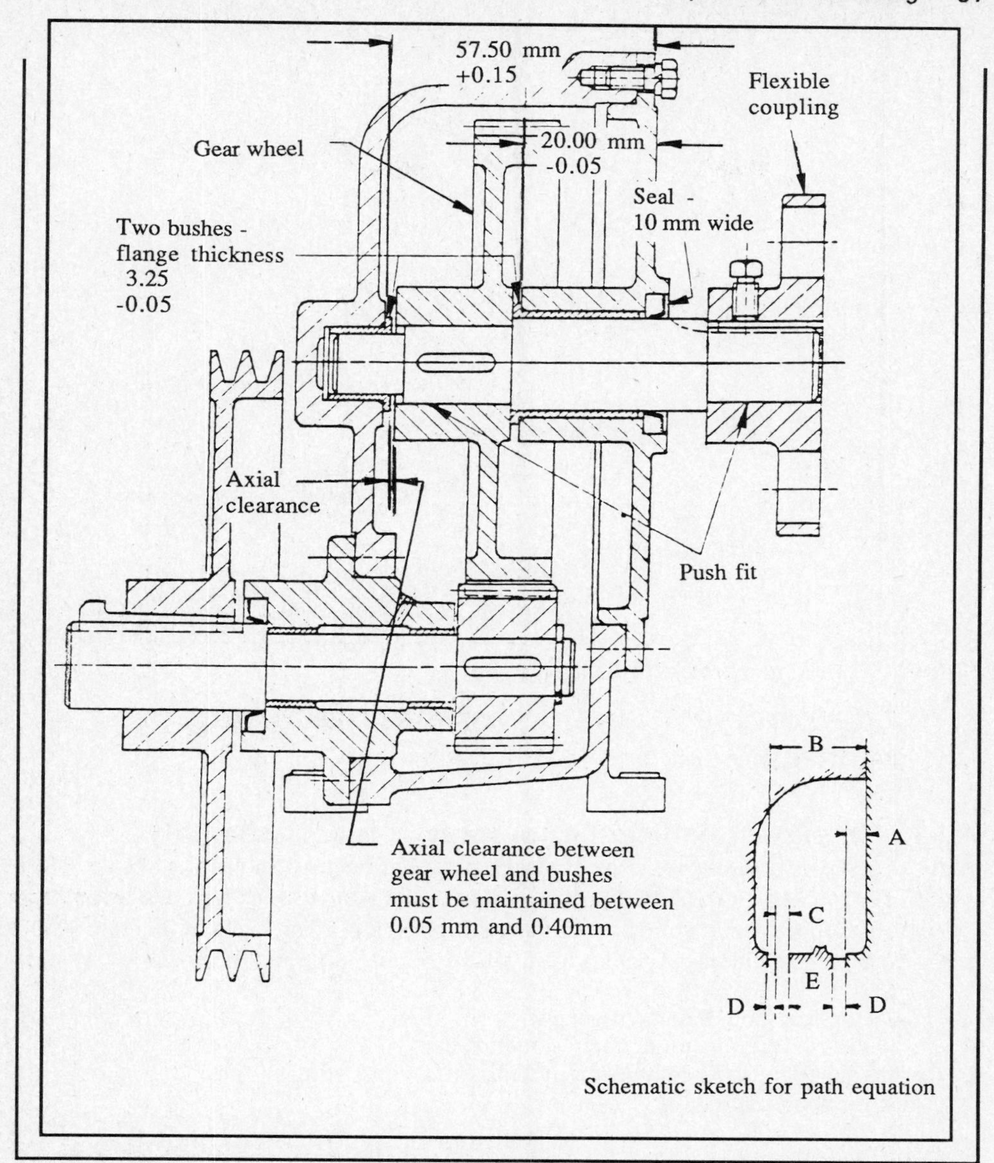

Figure 7.4 Sectional view of gear box

Solution

$$\text{Boss length} = 30.95 {}^{+0.00}_{-0.05} \text{ mm}$$

Isometric sketch of gearbox casing is shown in Figure 7.5

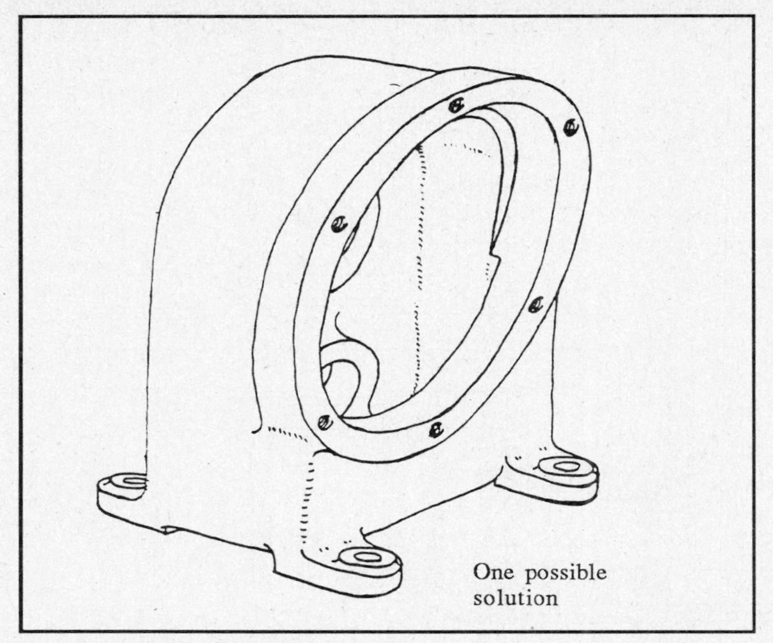

Figure 7.5 Isometric sketch of gearbox casing

7.7 An exercise in design for manufacturing variability

7.7.1 Design specification for relay contact assembly

Figure 7.6 shows a miniature relay contact assembly drawn to an enlarged scale. The center-lines marked 1 and 2 are the center-lines of the bolts used to hold the assembly together. The functional requirement of the assembly is that a gap width of 0.5 mm $\pm$ 50 μ is to be achieved without any adjustment after assembly. (1μ = 0.001 mm.)

A = contactor size (both flat and rounded)
B = insulator used for separating metal components
C = spring material used for contactor mounting and conducting current
D = spring support plates

1. *Write down* the path equation for the gap width, that is the equation which expresses gap width as a function of the other relevant dimensions in the assembly. You may assume that the first estimates for thickness of B, C and D are:

   B = 1.5 mm
   C = 0.4 mm
   D = 0.75 mm

2. *Determine* the dimension A of the contacts and allocate tolerances to all components to meet the functional requirements.
3. If possible, *adjust* component dimensions and tolerances to reduce manufacturing

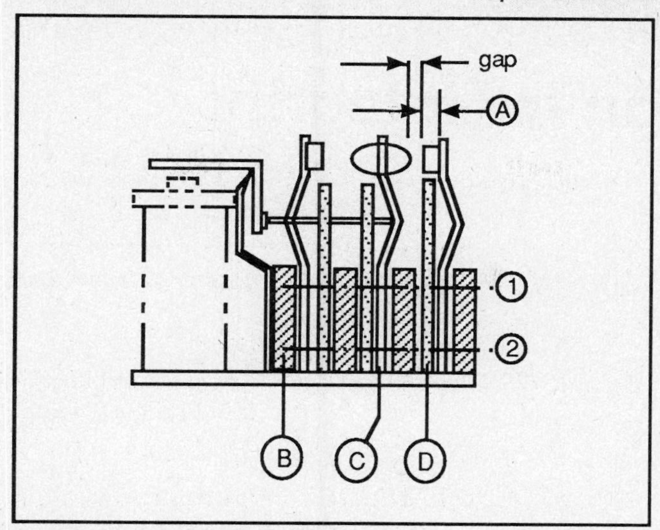

Figure 7.6 Schematic view of relay contactor assembly

costs.

Note: It is not necessary to understand how the relay works in order to answer this question.

Chapter 8

Design for integrity

When things can't get worse, they do.
Murphy.

Concepts introduced	theories of failure; factors of safety.
Methods presented	design method I.D.E.A.S.; identification of failure.
Applications	(a) tensile testing gripper for concrete test specimens; (b) laboratory exhaust chimney.

8.1 Design procedures

8.1.1 Structural integrity: Design against failure

We consider the design of common elements of engineering equipment such as machines and structures. We are concerned with applications where the successful performance of the equipment for a specified period of time in a specified environment depends on the stresses and strains to which its elements are subjected. Our attention is now directed towards *the lowest level of the design hierarchy* (Section 1.4) on the principle that a chain is as strong as its weakest link. It will be appreciated that we are looking at one particular class of design problem. To put this work in to perspective against the other activities of the engineering designer, reference should be made to the earlier chapters of this book, as well as to a useful publication by the British Standards Institution, P.D. 6112, *Guide to the Preparation of Specifications of Engineering Equipment.*

In the context of our discussion, the designer's responsibility is to choose the material and dimensions of an element to ensure that it has sufficient '*strength*' not to '*fail*' under the '*loads*' which it experiences in service, or to ensure that the probability of such a failure is reduced to an acceptably low value. The discussion will be at a fairly elementary level and limited to common materials of construction. In practice the engineer encounters many different materials — metals, alloys, timber, concrete, ceramics, polymers, rocks, soils, and so on — and has to be conversant with a wide variety of properties.

Under load, most engineering materials behave in a *ductile* or *brittle* manner depending on whether or not they exhibit significant deformation before fracture. Also various *stress-strain relationships* are observed.

Our discussions will focus on the most commonly occurring case where the material of construction is ductile and deforms elastically under load. (Mild steel satisfies these conditions at ordinary temperatures.) It is often *reasonable to assume* that the material is *homogeneous* and *isotropic*.

An engineering element constructed from such a material is usually unable to perform its intended structural function once a significant amount of yielding or plastic deformation has occurred. (There may be other ways in which the element can fail and which require separate consideration by the designer — see Section 8.2.) This observation leads to two alternative design procedures, one based on preventing the onset of local yielding, the other based on the overall load capacity of the element.

The first approach is commonly used in the design of machinery and equipment where even a small amount of local yielding in an element may lead to its catastrophic failure. The approach is based on the concept of an 'allowable stress', which is a fraction of the yield strength obtained by applying a 'stress factor' to give a suitable margin of safety. It is assumed that the material behaves elastically and that the onset of local yielding sets the limit to the usefulness of the element. The loads which the element experiences in service are used in the design calculations.

The second approach is widely used in structural engineering where it leads to methods of 'plastic design'. A beam or other structural element may exhibit significant plastic deformation across a critical section and still be capable of successfully resisting the applied loads. (See B. Gorenc and R. Tinyou, 1985.)

In plastic design procedures :

1. ultimate loads are used, these being obtained by multiplying the service loads by appropriate load factors;
2. forces and moments in structural elements are calculated allowing for inelastic behaviour;
3. structural elements are so proportioned that their strength is greater than or equal to the forces and moments produced by the ultimate loads.

According to the design procedure used, a *margin of safety* is provided by a *stress factor* or a *load factor*. This is further discussed in Section 8.4 under the heading of Factors of Safety.

8.1.2 A note on design for failure

In our concern for strength and reliability and durability, we should not overlook the fact that there are devices where the material of construction is *designed to fail* as:

1. an essential part of normal operation; for example the tear-away ends of beverage and soft drink cans; staples; adhesive tape;
2. a safety feature to prevent a system malfunction leading to a catastrophe; for example shear pins which protect a shaft from excessive torque; bursting discs which rupture when there is a sudden, excessive pressure rise in chemical reaction vessel or other piece of equipment subject to internal fluid pressure.These devices are generally characterized as *mechanical fuses*.

These few examples should clarify the nature of these applications, whose further elucidation is, however, outside the scope of this book.

8.2 Definition of failure

8.2.1 General

An engineering element may fail suddenly and without warning (e.g. by fracture) with possibly catastrophic results, or it may fail gradually (e.g. by increasing deformation under applied load). In either case failure is associated with excessive stress or strain or with unstable structural behaviour.

An engineer may have to design an element to resist one or more of the following *modes of failure*:

1. fracture
   (a) static loading;
   (b) dynamic loading — fatigue;
   (c) impact loading;
   (d) stress corrosion;
   (e) creep.
2. excessive deflection
   (a) elastic (linear);
   (b) plastic (non-linear).
3. instability
4. wear and surface damage

The region or regions where the element will fail must be clearly identified by the engineer as part of the initial appreciation of the design problem.

8.2.2 Examples of failure

In the following examples 1 to 5, P_{cr} is the critical value of the applied load which gives rise to a particular mode of failure of the element concerned. P_{cr} is thus a measure of the strength of that element.

1. fracture of bar of uniform cross-section (*area* A) subject to steady axial tensile force, brittle material of ultimate tensile strength S_u:
$$P_{cr} = S_u A$$
2. fracture of bar of uniform cross-section (*area* A) subject to a *reversed axial load* which varies sinusoidally with time, common ferrous material of endurance limit S_e:
$$P_{cr} = S_e A$$
3. excessive deflection (elastic) of simply supported beam of span l, second moment of area I, under central transverse load P, material of Young's modulus E:
$$P_{cr} = \frac{48\,EI}{l^2}\left(\frac{\delta}{l}\right)_{cr}$$
4. excessive deformation (yielding) of bar of uniform cross-section (*area* A) subject to steady axial force, ductile material of yield strength S_y:
$$P_{cr} = S_y A$$
5. buckling of slender column of effective length l:
$$P_{cr} = \frac{\pi^2 EI}{l^2}$$

8.3 Prediction of failure

Theory from mechanics of solids together with a knowledge of the properties of engineering materials is used to predict the strength of an element corresponding to a particular mode of failure.

If the mode of failure is yielding or fracture and the material in the element is subject to biaxial or triaxial stressing, then a difficulty arises. Information on the strengths of engineering materials (properties such as S_y, S_u, S_e) is obtained in standardized laboratory tests in which the stressing is one-dimensional (usually tension). Some connecting link is required between the data from one-dimensional tests and the behaviour of the material under biaxial or triaxial stress. To predict failure by yielding or fracture under biaxial or triaxial stress a theory of failure is necessary.

This is, strictly speaking, an hypothesis based on plausible reasoning which incorporates some function of stress and strain to link one-dimensional and three-dimensional material behaviour.

Results from the theory of mechanics of solids will be used. We consider an element of material subject to triaxial stressing (refer to Figure 8.1); the material is homogeneous, isotropic, elastic. There will be three principal stresses, σ_1, σ_2, σ_3, acting on mutually perpendicular planes on which the shear stresses are zero.

The usual convention is adopted that $\sigma_1 > \sigma_2 > \sigma_3$, so the maximum shear stress is of magnitude:

$$|\tau_{max}| = \frac{1}{2}(\sigma_1 - \sigma_3)$$

The shear strain energy per unit volume is:

$$U = \left(\frac{1 + \mu}{6E}\right)\{(\sigma_1 - \sigma_2)^2 + (\sigma_2 - \sigma_3)^2 + (\sigma_3 - \sigma_1)^2\}$$

To predict failure (either by fracture or by onset of yielding) under combined stresses,

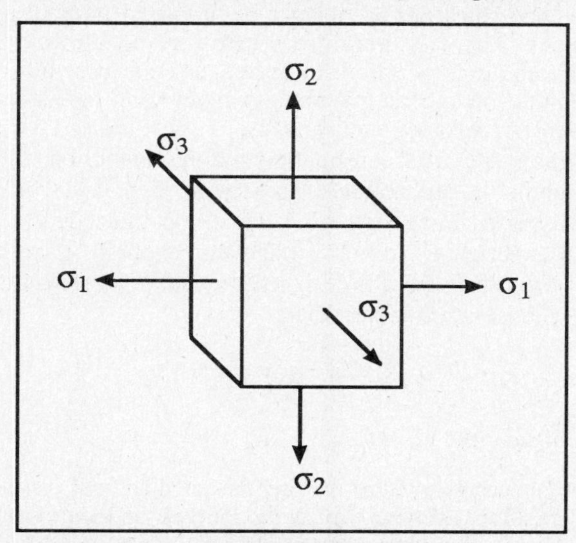

Figure 8.1 An element of material subject to triaxial stress

the following theories of failure have been put forward. Their relevant applications are shown in brackets. For the time being steady loads only will be considered.

1. *Maximum principal stress theory (fracture of brittle materials)*
   This theory predicts that failure (fracture) will occur when the maximum principal stress in the stressed material reaches the value of the tensile strength in a simple tension test. In symbols, *we predict fracture* when:
   $$\sigma_1 = S_u$$
2. *Maximum shear stress theory (yielding of ductile materials — theory is conservative)*
   This theory predicts that failure (yielding) will occur when the maximum shear stress in the stressed material equals the shear stress corresponding to the yield strength in simple tension. In symbols, we predict yielding when:
   $$\tau_{max} = \frac{1}{2} S_y;$$

that is, $$(\sigma_1 - \sigma_3) = S_y$$

3. *Maximum shear strain energy theory (yielding of ductile materials — more accurate than (2) but not as conservative)*
   This theory predicts that failure (yielding) will occur when the shear strain energy per unit volume in the stressed material is equal to the shear strain energy per unit volume in a tension test specimen at yielding. In symbols, we predict yielding when:
   $$(\sigma_1 - \sigma_2)^2 + (\sigma_2 - \sigma_3)^2 + (\sigma_3 - \sigma_1)^2 = 2S_y$$

8.4 Factors of safety

8.4.1 Introduction

The failure of an engineering element of the type discussed in these notes usually has serious consequences. There is therefore a heavy responsibility on designers to incorporate adequate margins of safety to ensure that the possibility of a failure is extremely remote. On the other hand they want to avoid using such an excessive amount of material that the design becomes unduly costly.

In deciding on the material of construction and the dimensions of an element, the designer has to weigh up the probabilities and exercise judgment. The designer may do this explicitly (as discussed in Section 3.3.2, Ideas and Concepts), or implicitly from previous successful experience. To make judgments explicit he/she may introduce a *Factor of Safety* (F_d) to ensure that the designed strength of an element (S_d) exceeds the load applied to it (L_d) by a safe amount, where:

$$F_d = \frac{S_d}{L_d}; \quad \text{with } F_d > 1$$

The design *margin of safety* is: $$M_d = S_d - L_d$$

If a complete engineering structure is being designed and not just one element, then a *Load Factor* is defined in a similar way as the ratio of the load to cause failure of the structure to the working load used as a basis for design.

8.4.2 Implicit factors of safety

The designer is likely to extrapolate from successful previous experience when the parts being designed are of complicated geometry and/or are subject to complex loadings varying perhaps in both space and time. In such situations it is often difficult if not impossible to apply mechanics of solids theory; it may not even be possible to identify all the relevant modes of failure. Examples are common in transport vehicle design (e.g. stub axles of cars and trucks), and for this reason the automotive industry uses special proving grounds for testing and verifying the structural integrity of new cars and trucks. Even so, there may be uncertainty as to how well accelerated life testing on proving grounds predicts the lives of automotive components in the wide variety of physical environments a car is exposed to over its working life.

8.4.3 Explicit factors of safety

Investigation of the working methods of professional engineers shows that they frequently rely on qualitative judgments (Lewis, 1985). Examples: that a proposed course of action has 'a good chance' of success, or alternatively is 'a poor risk'. Other phrases frequently used are: 'a wild chance', 'hairy', 'a good bet' and so on, much the same as in everyday affairs.

Nevertheless, in certain circumstances it becomes necessary to estimate an explicit factor of safety and convert such qualitative judgments into quantitative estimates:

1. when the element being designed is novel so that there are few or no precedents for predicting its successful performance;
2. when the element being designed would require large resources for its implementation so that economy of material is a major consideration;
3. when existing precedents or codes of practice contain conflicting recommendations.

Various groups of engineers have tackled the problem of dealing with uncertainties in design. The ensuing discussion has been influenced by the proceedings of specialist conferences (Rossmanith, 1984; Wen, 1984) and by the work of Black (1972) and Pugsley (1966). We consider the design of a typical engineering element which is in turn part of some larger structure, and examine the sources of:

1. uncertainty regarding the load applied to the element;
2. uncertainty regarding the predicted strength of the element.

Each relevant source of uncertainty has to be allowed for in the *factor of safety* adopted (see Table 8.1).

In addition, a further factor (F_O) is introduced to take into account the consequences of failure. *'It is in the nature of the human mind to require a margin of safety measuring up to the seriousness of the consequences of failure.'* (Pugsley, 1966)

Each of the above factors acts independently of the others so that the *design factor of safety* (F_d) is obtained by multiplying these individual factors together.

Note: (a) Each $l > 1$

(b) Each $s > 1$; *if it acts to reduce* the strength of the element.

The following guide is given for estimating F_d. In this approach it is important to ensure that systematic errors are not introduced as, for example, by a tendency to assess each factor conservatively.

$$F_d = F_O\, l_1\, l_2\, l_3\, s_1\, s_2\, s_3\, s_4\, s_4\, s_5$$

Table 8.1 Sources of uncertainty and factors of safety

		Source of uncertainty	*Factor*
1.	(a)	Magnitude of load applied to structure	l_1
	(b)	Rate at which load is applied	l_2
	(c)	Sharing a load between elements of structure	l_3
2.	(a)	Variation in material properties	s_1
	(b)	Effects of manufacturing process (Is there a greater probability of defects being present in element than in laboratory specimens, or vice versa?)	s_2
	(c)	Effects of physical environment (corrosion, erosion, temperature)	s_3
	(d)	Effects of stress concentrations	s_4
	(e)	Assumptions made in constructing mathematical model on which design predictions are based	s_5

l_1 l_2 l_3 s_1 s_2 s_3 s_4 s_4 s_5: depend on accuracy of analysis, knowledge of materials, knowledge of loads applied and strength to resist the loads. (l_2 varies from 1.2 — *light shock* — to 3 or more for *heavy shock*.)

Subjective rating	*Factor*
'very good'	1.1
'good'	1.3
'fair'	1.5
'poor'	1.6

s_4: See Chapter 10 for discussion of stress concentrations.
F_0: Depends on consequences of failure.

For failure rated:		
	'very serious'	$F_0 = 1.4$
	'serious'	$F_0 = 1.2$
	'not serious'	$F_0 = 1.0$

8.4.4 Probabilities

One school of thought favours the explicit use of probabilities; see for example Freudenthal (1969), Wen (1984). The argument runs as follows. As before, we recognize that there are uncertainties in the design of an engineering element because of our lack of detailed knowledge of the load imposed on it and of its strength in resisting that load. Instead of considering one (more or less typical) value of the load applied to the element, we consider the set of all possible values of this load and endow this set with a probability density function $f(x)$. Suppose that the set of possible values of the load is the set (0 to ∞). Then we assume that $f(x)$ will satisfy the normalizing condition.

$$\int_0^\infty f(x)dx = 1$$

and we take the integral $\int_a^b f(x)dx$ as a measure of the probability of the load L satisfying the inequality $a < L < b$.

Note that the probability measure thus defined does not necessarily refer to a relative frequency of occurrence of the event $a < L < b$, since it may well be that the structure or machine under consideration is unique. It is rather a numerical expression of accumulated engineering experience and judgment. The designer uses own experience, that of other designers, and the information available from previous designs.

We also consider the set of all possible strengths of the element and endow this with a probability density function $g(x)$ with:

$$\int_0^\infty g(x)dx = 1$$

The probability of failure is then: $$\int_0^\infty f(x) \{ \int_0^x g(x)dx \} dx$$

Suppose that the penalty for failure is V, expressed in monetary terms and covering cost of repair, compensation for damage to life and property, and compensation for damage to reputation. Suppose also that the cost of producing a successful design is C, then ideally the designer should minimize the expected total cost $E(T_C)$, where:

$$E(T_C) = C + pV$$

However, this ideal approach is remote from day-to-day practice. A less refined probabilistic method is to specify probability of failure arbitrarily, taking into account only the social acceptability of the risks involved. The best design is then that which minimizes C subject to the constraint $p < (p)_{specified.}$

8.5 Design methods

8.5.1 Design from first principles: I.D.E.A.S.

We are now in a position to state formally the procedure for designing elements of engineering machines and structures. Our notation is as follows:

S_i = Strength of element in i^{th} mode of failure.
L_i = Load causing i^{th} mode of failure of element.
F_i = Factor of Safety for i^{th} mode of failure.

The design procedure consists of five steps:

1. **I**dentify modes of failure.
2. **D**erive equations predicting strength of element, that is, load at which it fails in each mode.
3. **E**stimate factor of safety for each mode of failure.
4. **A**pply design inequalities, $S_i > F_i L_i$.
5. **S**elect material and dimensions of element so that all inequalities are satisfied.

Notes:

1. The initial letters of steps 1 to 5 form a convenient mnemonic:

   I.D.E.A.S.

2. Step 5 often involves trial and error and the designer should be prepared to carry out several iterations to obtain an acceptable outcome.
3. If the element being designed is a costly item then step 5 should be carried further to ensure that an optimum minimum cost design is achieved.
4. Selection of material. The examples presented in Section 8.2.2 were of the form $S = GP$, where S is the strength of the element, G is a geometrical factor, a function of the dimensions of the element and P is a material property. This pattern exists is many design problems and provides the basis for material selection by determining an appropriate value for P.
5. The design is implemented in accordance with the general procedures stated in Chapter 7.

8.5.2 Design from precedent

The design of new products, artefacts, structures and machines requires the ability to work from first principles and apply the five-step procedure described above. But designer time, like most other resources, is usually in short supply, and recourse may well be had to successful precedents where established methods have been demonstrated to be time-efficient and to lead to acceptable results. In many countries the repertoire of successful precedents is encapsulated in Standard codes of practice, to the point where the proliferation of such Standards is often confusing to the young engineer. Table 8.2 has been constructed to display those Standards from Britain, America, and Australia most relevant to the design of elements of structures and machines for structural integrity. Space does not permit a survey of the multitude of Standards which might be encountered in practice: the data in the Table cover common applications where design thinking is assisted by recourse to precedent.

For the professional engineer, Standards represent a codification of information on safe, effective design procedures and on appropriate materials of construction. For the student engineer, Standards provide a mental crutch to support thinking about problems which might otherwise appear to lack clearly defined starting points or methods of solution. However, in the educational context a word of warning is in order. The application of Standard procedures and methods in practice is leavened with the judgment born of experience. Students reading this text need to be cautioned to develop a critical faculty, a questioning attitude which does not blindly accept the authority of a Standard, even one pre-eminent in its field, without assessment of the theoretical ideas and experimental evidence on which it is based.

The examples and problems in Chapters 9 to 14 blend theory, first principles, and engineering judgment in an educational amalgam to promote the transition from novice to accomplished practitioner.

Table 8.2 Design standards for common engineering applications

	Australian Standard	*British Standard*	*United States Standard*
Dimensioning and tolerancing	AS 1100	BS 308	ANSI Y 14.5M
Specifications of common steels	AS 1142 AS 1483	BS 970	ASTM A283 et al see ASTM publications
Methods of fatigue testing	—	BS 3518	ASTM E466, E467, E468, and E739
Design of steel structures	AS 1250	BS 5950	AISC specification, manual of steel construction
Design of pressure vessels	AS 1210	BS 5500	ASME boiler and pressure vessel code
Design of shafts for power transmission	AS 1403	—	ANSI B 106.1M
Strength of screw threads	AS B 232	BS 3580	—

AISC = American Institute of Steel Construction
ANSI = American National Standards Institute
ASME = American Society of Mechanical Engineers
ASTM = American Society for Testing Materials

8.6 Notes from a designer's workbook

8.6.1 Design of a tensile testing gripper for concrete test specimens

Initial information (general layout, as shown in Figure 8.2, adapted from Harvey, 1965):

1. Concrete specimens are 150 mm square section by 900 mm long.
2. Tensile strength of concrete is about 3.5 to 5 MPa.
3. The average value of the friction coefficient between steel and concrete is 0.35.

Design problem

1. Show that the mass of the loading attachment is given by:

$$M = \frac{A\rho}{\sigma_t \sin 2\theta} + \frac{B\rho}{\sigma_t \sin 2\alpha} + c\,t\,\rho \tan\alpha + (2M_r + M_b) \tag{8.1}$$

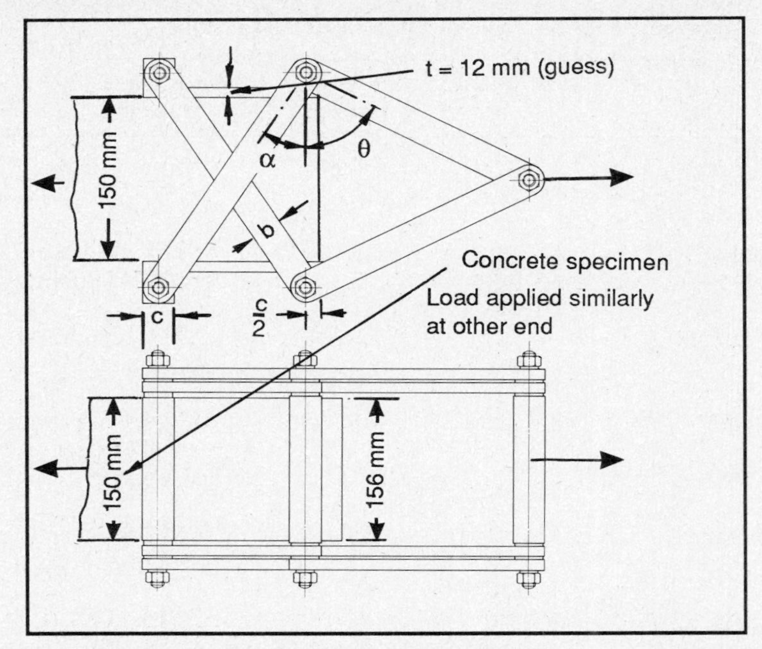

Figure 8.2 Concrete test specimen gripper (schematic)

where: $$\tan\alpha = \frac{2\,\mu\,\tan\theta}{\tan\theta - \mu}$$

$A = 2.25 \times 10^4$; $B = 4.5 \times 10^4$; $C = .062$.
ρ = density of steel.
σ_t = allowable tensile strength of links (neglecting end effects).
μ = coefficient of friction between steel side plates and concrete.
M_b = mass of loading bar.
M_r = mass of roller.

At this early stage it may be assumed that M_r is insensitive to changes in θ (check later). With $\rho = 7800$ kg/m^3; t = 12 mm ; $\mu = 0.35$, the equation (8.1) becomes:

$$M = \frac{175.7 \times 10^6}{\sigma_t \sin 2\theta} + \frac{351.3 \times 10^6}{\sigma_t \sin 2\alpha} + 5.84 \tan\alpha + K \text{ (constant)} \qquad \mathbf{(8.2)}$$

2. Plot [M – K] against θ for values of σ_t of 70, 104 and 140 MPa, and use the plot to determine the optimum value of θ.
3. Design rollers for minimum weight. In other words, determine roller diameter and material.
   Note that the rollers are subject to bending as well as contact stresses.
4. Make a sketch of the roller, showing all information necessary for manufacture.

Additional information:

1. From mechanics of solids theory, contact stress $\tau_c = 0.419\sqrt{\frac{p' E}{R}}$

   where: p' = load per unit length;
   E = Young's modulus of elasticity;
   R = roller radius.
2. Factors of safety (from experience):
   Bending F_d = 4
   Contact F_d = 1.25

Solution

1. Force analysis

(a) equilibrium of loading pin : $2P_1 \sin\theta = \frac{p}{2}$; $P_1 = \frac{p}{4 \sin\theta}$

(b) equilibrium of roller:
Assume friction between roller and side plate is negligible; hence R_1 is very nearly perpendicular to plate.

$P_1 \sin\theta = P_2 \sin\alpha$; $P_2 = \frac{p}{4 \sin\theta}$; $R_1 = 2P_1\cos\theta + 2P_2\cos\alpha$
$= \frac{p}{2}(\cot\theta + \cot\alpha)$

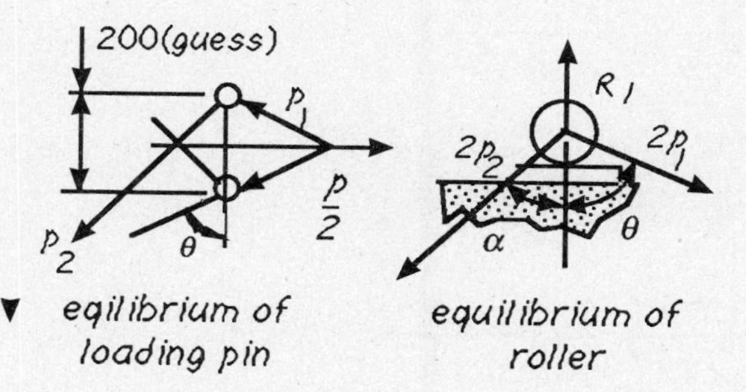

equilibrium of loading pin equilibrium of roller

(c) equilibrium of left-hand pin: $R_2 = 2P_2\cos\alpha = \frac{p}{2}\cot\alpha$

(d) equilibrium of gripper: when slipping is about to occur, $2\mu R_1 + 2\mu R_2 = p$; substituting from above gives:

$$\tan\alpha = \frac{2\mu \tan\theta}{\tan\theta - \mu}$$

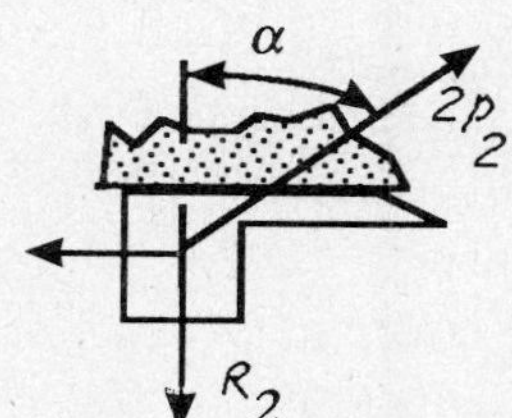

Equilibrium of left-hand pin

2. Mass of loading attachment

(a) member (1)

length $l_1 = \dfrac{x}{2\cos\theta}$; area $A_1 = \dfrac{p_1}{\sigma_t}$;

$$\text{mass } m_1 = \rho \times A_1 \times l_1 = \frac{\rho\ p\ x}{4\ \sigma_t\ \sin 2\theta}$$

(b) member (2)

length $l_2 = \dfrac{x}{\cos\alpha}$; area $A_2 = \dfrac{p_2}{\sigma_t}$;

$$\text{mass } m_1 = \rho \times A_1 \times l_1 = \frac{\rho\ p\ x}{4\ \sigma_t\ \sin 2\theta}$$

(c) side plate (s)

length $l_s = x \tan\alpha + \text{constant}$; area $A_s = w\ t$;

$$\text{mass } m_s = \rho \times A_s \times l_s = \rho w\ tx \tan\alpha + \text{constant}$$

(d) total mass

$$M = 4m_1 + 4m_2 + 2m_s + 2m_r + m_b$$

$$= \frac{\rho\ p\ x}{\sigma_t\ \sin 2\theta} + \frac{2\rho\ p\ x}{\sigma_t\ \sin 2\alpha} + 2\rho\ w\ tx \tan\alpha + k\ (\text{constant})$$

substituting:

p = 112.5kN (specimen area x maximum tensile load);

ρ = 7800 kg/m^3; x 0.2m; t = 0.012m; w = 0.156;

μ = 0.35;

$$M - K = \frac{175.7\times10^6}{\sigma_t\ \sin 2\theta} + \frac{351.3\times10^6}{\sigma_t\ \sin 2\alpha} + 5.84 \tan\alpha$$

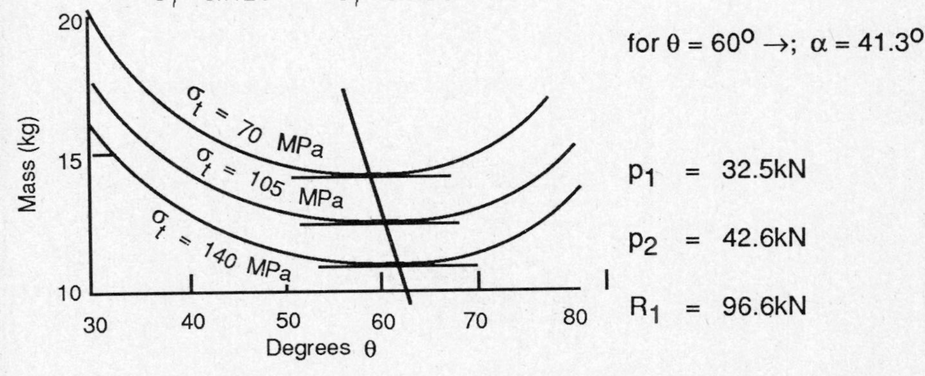

M is plotted against θ for various values of σ_t . As can be seen, θ_{opt} is not very sensitive to changes in σ_t . Also the change of mass is small within the range of five degrees either side of optimum θ. Thus we can postpone the decision on the material to be used. A suitable choice of θ would be 60°

3. Design of roller
(a) Bending stresses
Consider roller as a uniformly loaded, simply supported beam.

$$\text{load} = \frac{R_1}{0.156} = \frac{96.6}{0.156} = 619.2\ \text{kN/m};$$

$$\text{end reactions} = \frac{96.6}{2} = 48.3\ \text{kN}$$

$$M_{max} = 48.3 \times \left(\frac{.16}{2} - \frac{.156}{2}\right) = 1.884\ \text{kNm}$$

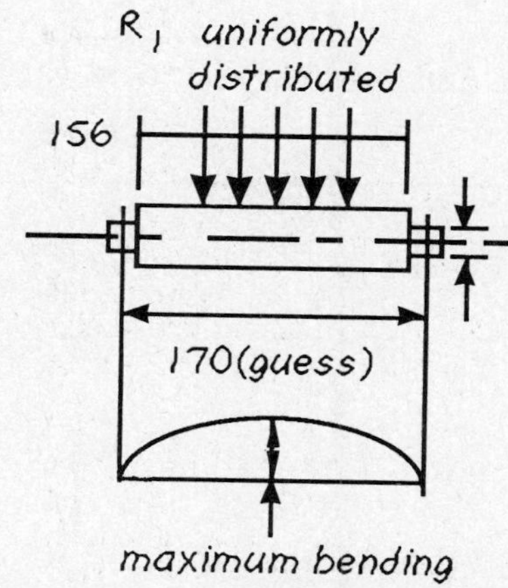

Assume ductile material and use maximum shear theory of failure:

yielding occurs when the maximum shear stress reaches $\frac{S_y}{2}$;

design with factor of safety $f_d = 4$;

allowable shear stress $\tau_{all} = \frac{S_y}{2f_d} = \frac{S_y}{8}$;

design rule : $\tau \leq \tau_{all}$; at the roller surface in the central plane

we have : $\tau = \frac{1}{2} \frac{M_{max}}{\frac{\pi d^3}{32}}$; uniaxial stress distribution gives:

$$\tau = \frac{\sigma}{2} = \frac{My}{2I_{zz}};$$

thus $\tau = \frac{16}{\pi} \frac{1.884 \times 10^3}{d^3} \leq \frac{S_y}{8}$ which gives:

$$\boxed{d^3 \geq \frac{76.8}{S_y}}$$

(b) Contact stresses

These stresses are quite independent of the bending stresses calculated above.

As previously noted
$$\tau_c = 0.419\sqrt{\frac{p'\,E}{R}}$$
$$= 0.419\sqrt{\frac{619.2\times10^3\times210\times10^9\times2}{d}}$$

and this must be less than $\frac{S_y}{2\times1.25}$; that is $\frac{2.137\times10^8}{\sqrt{d}} \leq \frac{S_y}{2.5}$;

this gives another design inequality for the roller diameter d;

$$\boxed{d \geq \frac{2.854\times10^{17}}{S_y^2}}$$

These two inequalities can be plotted to establish an optimum value for the roller diameter.

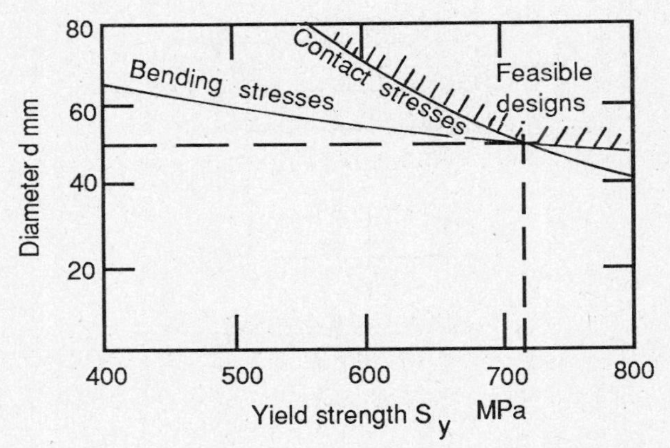

The optimum diameter is 50 mm using a steel of 720 MPa yield strength.

8.6.2 A laboratory chimney

You are required to design a chimney of height 9 m. The chimney is a vertical tube used to discharge small quantities of radioactive gas into the atmosphere. It will be mounted on the roof of a research laboratory and for aesthetic reasons guy ropes or wires will not be used to help support the chimney. It has been decided to construct the chimney from galvanized mild steel and for simplicity of design, a tube of constant cross section will be used.

The dimensions of commercially available tubing are given in Table 8.3.

Two safety decisions have been taken from the appropriate standards:

1. the maximum tensile or compressive stress at the bottom of the chimney will not exceed 21 MPa; and
2. the maximum deflection at the top of the chimney in an 80 km/h wind must not exceed 25 mm.

Table 8.3 Dimensions of commercially available tubing

Outside diameter (mm)	*Tube thickness (mm)*
125	5
140	6
150	5;6
165	5;6;8
190	5;8;10

The design problem

1. List the possible modes of failure of the chimney.
2. Investigate the individual factors affecting the selection of the design factor of safety F_d.
3. Design the chimney to meet the stress and deflection safety requirements. Try to choose the 'best' tube section for the task.
   In your calculation you may assume that the aerodynamic loading in an 80 km/h wind on a circular cylinder of uniform diameter D mm is given by:
   $$f = 0.31D \text{ N/m}$$
4. Estimate the effect on stress and deflection of a small error (say 2 percent) in the height of the chimney.

Solutions

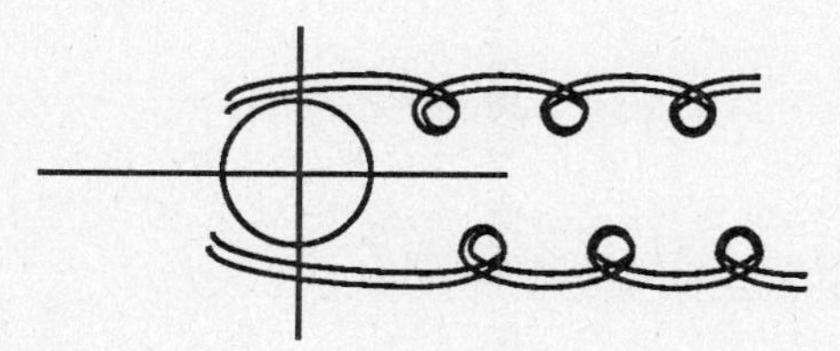

Vortex shedding from a cylinder placed in a fluid stream

1. Modes of failure
Assume that the chimney behaves like a vertical cantilever.
There are three main modes of failure to be considered:

(a) yielding at the base due to direct wind loading;
(b) excessive deflection under wind loads;
(c) yielding or fracture at the base of the chimney due to dynamic loading.

With respect to (c), it is noted that vortex shedding from the cylindrical chimney will impose lateral dynamic loads on it as the vortices are shed from alternate sides of the cylinder in a 'vortex street', as shown on the sketch.

If the rate of vortex shedding is near the natural frequency of vibration of the chimney, large vibrations will be set up, resulting in large stresses and possibly yielding or fatigue fracture at the base of the chimney. In the calculations below this third mode of failure is ignored, and consequently a large factor of safety is necessary.

2. factor of safety

$$f_d = f_c \, [\, l_1 l_2 l_3 \; s_1 s_2 s_3 s_4 s_5 \,]$$

The design factor of safety is f_d estimated below:

l_1 =	1.5 ;	our knowledge of load magnitude is only fair. The figure of 0.31D N/m is well established but winds higher than 80 Km/h may be encounterd in practice.
l_2 =	3.0 ;	chimney is designed for static loading only – dynamic effects are ignored.
l_3		not applicable.
s_1 =	1.3 ;	material properties well known but some variations may be expected in commercial mild steel.
s_2 =	1.0 ;	influences of manufacturing process would be very small.
s_3 =	1.1 ;	steel protected against atmospheric corrosion.
s_4 =	1.1 ;	effects of stress concentration would depend on how the chimney was supported at its base. With good design, the effects of stress concentration will be small.
s_5 =	1.1 ;	at ordinary temperatures, > 15°C, mild steel behaves elastically and the theory of elasticity can be applied with confidence. If temperatures were low enough to cause brittle behaviour in steel, then a separate investigation would be necessary. It would not be acceptable to simply increase S_5.
f_c =	1.2 ;	failure is serious because small amounts of radioactive gas may be released into the atmosphere where they would not be dispersed by wind. Also falling chimney may damage people or property.

$$fd = 1.2(1.5 \times 3.0) \times (1.3 \times 1.1^3) = 9.2$$

The 21 MPa safe stress implies a factor of 10 for mild steel.

3. Design of Chimney

deflection : $\frac{ql^4}{8EI} \leq 25 \times 10^{-3}$

where $q = 0.31\ D \times 10^3$ N/m;

$l = 9$ m;

$E = 210$ GPa;

D = diameter of chimney in meters.

$$I = \frac{\pi}{64}(D^4 - d^4)$$

$$\boxed{\frac{0.987\ D}{(D^4 - d^4)} \leq 10^3} \qquad (a)$$

(a) stress: axial stress negligible

bending stress: $\frac{My}{I} \leq 21 \times 10^6$

$M = \frac{ql^2}{2}$ Nm; $y = \frac{D}{2}$; I as above.

$$\boxed{\frac{6.09D^2}{(D^4 - d^4)} \leq 10^3} \qquad (b)$$

As a first approximation solve (a) and (b) simultaneously:

$D = 162$ mm $\qquad d = 150$ mm $\qquad$ 165 @ 6mm will do.

Now we must check other tubes near this value to compare performance.

The two design inequalities give:

$$\boxed{d^4 \leq \{D(D^3 - 0.987 \times 10^{-3})\}} \qquad (a)$$

$$\boxed{d \leq \sqrt{D(D^2 - 6.09 \times 10^{-3})}} \qquad (b)$$

These two inequalities may be plotted on a two-dimensional design space for d and D to find the least mass solution.
Alternatively the mass (proportional to $(D^2 - d^2)$) may be found for all suitable (D/d) combinations to find the smallest mass.

D	d	$\frac{0.987D}{(D^4 - d^4)}$	$\frac{6.9\ D^2}{(D^4 - d^4)}$	$(D^2 - d^2)$
150	138	1030	950	unsuitable
165	153	843	858	.0038 (OK)
165	155	993	1011	unsuitable
190	180	740	867	.0037 (OK)

190 x 5 mm wall thickness is best

(c) 2% error in length of chimney
shorter (-2%) → lower stress, lower deflection
longer (+2%) → deflection increased by factor 1.02^4 (i.e. +8%)
stress increased by factor 1.02^2 (i.e. +4%)

8.7 Exercises in design for structural integrity

8.7.1 Simple structure

Figure 8.3 shows a structure constructed from 10 mm x 20 mm mild steel and 10 mm diameter pins.

(a) List all the possible modes of failure for the structure under the action of the load P.
(b) Determine the load P at failure of the structure.
(c) Suggest ways of modifying the structure to improve the effectiveness of the use of materials.

The following material properties apply:

Mild steel structure: $S_u = 350\ MPa$; $S_y = 240\ MPa$
Pins: $S_s = 200\ MPa$; $S_b = 205\ MPa$
S_b is the allowable bearing stress (average contact pressure).

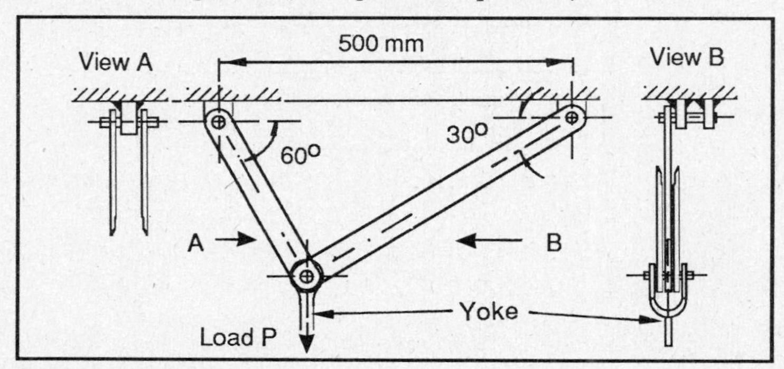

Figure 8.3 A simple structure
Note: Load P may be applied anywhere in the range shown.

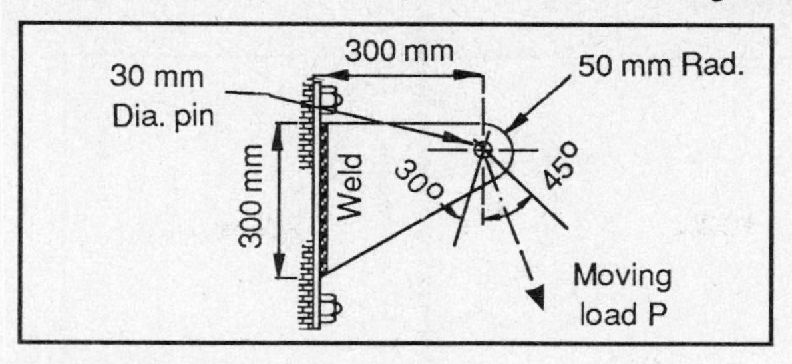

Figure 8.4 Mild steel welded bracket

8.7.2 Mild steel welded bracket

The Figure 8.4 shows a mild steel bracket designed to carry a load P. The load is due to a block and tackle supported from the bracket through the pinned joint passing through the hole shown in the bracket. The lifting gear is used to lift castings in a small scale brass founding shop.

(a) Estimate the factor of safety for each element of the bracket.
(b) Assuming that the weld is designed to carry as much load as the rest of the bracket, estimate the maximum safe load P. (Weld design not required in this exercise. See later in Chapter 13).

8.7.3 Oil-gas separator

You are required to design the main vessel of an oil-gas separator plant. The vessel is cylindrical, with hemispherical ends. This vessel is completely buried under ground except for one end, which is open to the atmosphere. Internal as well as external walls of the vessel have been treated to resist corrosion. The vessel is 275 m long, with an external diameter of 1.055 m. The internal pressure is 8.14 MPa.

Estimate the appropriate factor of safety to use in this design, given that the information based on standard pressure vessel design (and relevant data from the standard codes of practice) is of little value in cases where design experience is limited — as is the case with such oil-gas separators.

8.7.4 Timber stacking platform

Figure 8.5 shows a timber stacking platform used in a timber yard. Packs of timber, each weighing approximately 2 tonnes, will be placed on such platforms by a mobile crane (the crane's maximum safe load is designated at 3 tonnes). The platform is to be hinged along one edge, as shown, and it is supported by solid steel tie rods of circular cross section.

1. Estimate the factor of safety for this design.
2. Estimate the size of the tie rods.

8.7.5 Hydraulic pit-prop for mines

Hydraulic pit-prop for mines. Figures 8.6 and 8.7 show the construction and operating principles for a hydraulically operated pit-prop used in underground mining (McKee, 1964). The prop consists of a tubular column of medium carbon steel, having a

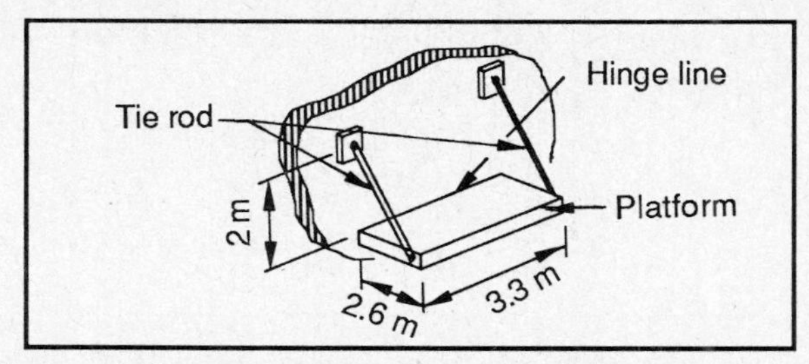

Figure 8.5 Timber platform schematic (dimensions in mm)

minimum yield strength of 315 MPa and a Young's modulus of 210 GPa. The outside diameter of the column must match the diameter of the actuator, namely 175 mm, and it is to be designed to carry an axial load of 405 kN. Each end of the prop rests on flat metal surfaces and the maximum height of prop required for a particular mining application is 4.3 meters.

Very few mines are located in solid, homogeneous bodies of rock and earth. The usual formations are layers or strata of rock, gravel and earth. These layers are tipped and often fractured. Such formations will tend to close up mine tunnels or other man-made holes, and specially designed props are required to resist these massive earth motions. The traditional solution is to use wooden props, which will 'yield' slightly, but will stay in place during the motion of earth above it. A wooden post, under compression, tends to spread, due to the buckling of the internal fibers, putting the outer fibers in tension. This behaviour of timber gave rise to the age-old method used by miners to hit the post with a hammer as a means of checking to see if 'the timber would fly'. If the splinters 'flew', this was a sign that the roof was moving down, even if very slowly.

In recent years adjustable metal props have come into wide use. A typical hydraulic prop design is shown in the following drawing. In its essentials it is similar in operation to an automobile jack, with a valve provided for pumping hydraulic oil, or to return it to a reservoir, at a predetermined pressure. This process provides a precise and reproducible yielding force or 'stiffness' of the prop. The prop is raised and pre-loaded from an external reservoir. The design shown also incorporates an accumulator charged with high-pressure inert gas, such as nitrogen for example, to give the required dynamic action to the prop. This is the ability to shorten or extend with small changes in force.

1. Draw the free body diagram for the component consisting of items 3 and 4 on the drawing, and show all the forces acting on it. Hence, determine the oil pressure in the actuator, when the compressive load on the column is 405 kN.
2. Check that the stresses in the cylindrical housing of the actuator, item 4 on the drawing, are within acceptable limits. The internal diameter of the housing is 145 mm. A detailed stress analysis is not required — just a quick check on the maximum stress, using simple theory.
3. Determine the thickness of the tubing to be used for the prop, given that it has already been decided to use a design factor of safety of 2.5. Commercially available tube thickness values are multiples of 2 mm. Justify the factor of safety suggested (2.5).
4. Suppose that commercially available tubing is manufactured to tolerances such that the outside diameter is $(D + dD)$ and the wall thickness is $(t + dt)$ where D and t are nominal values of the dimensions concerned:

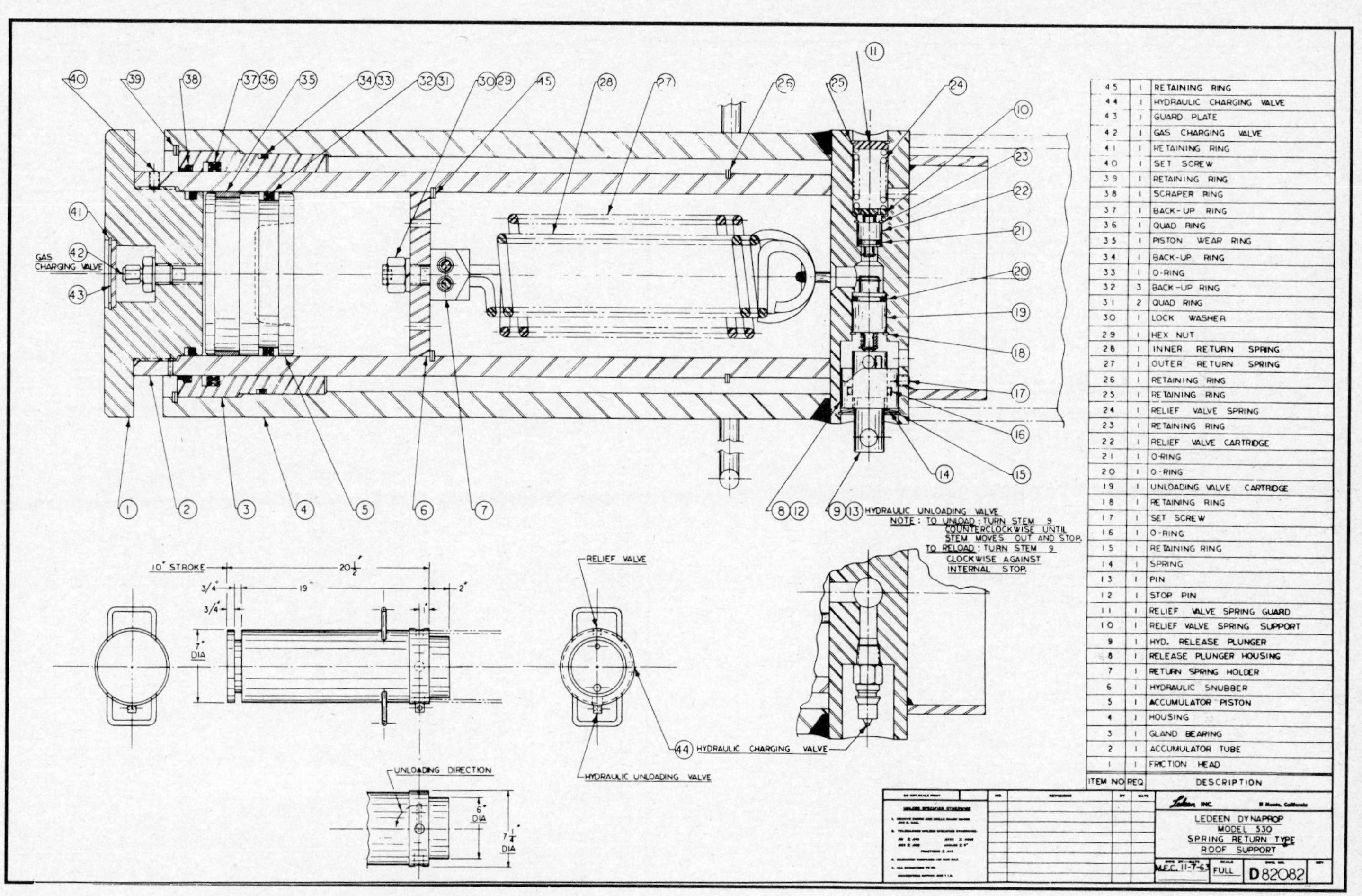

Figure 8.6 General layout drawing of a hydraulic pit-prop

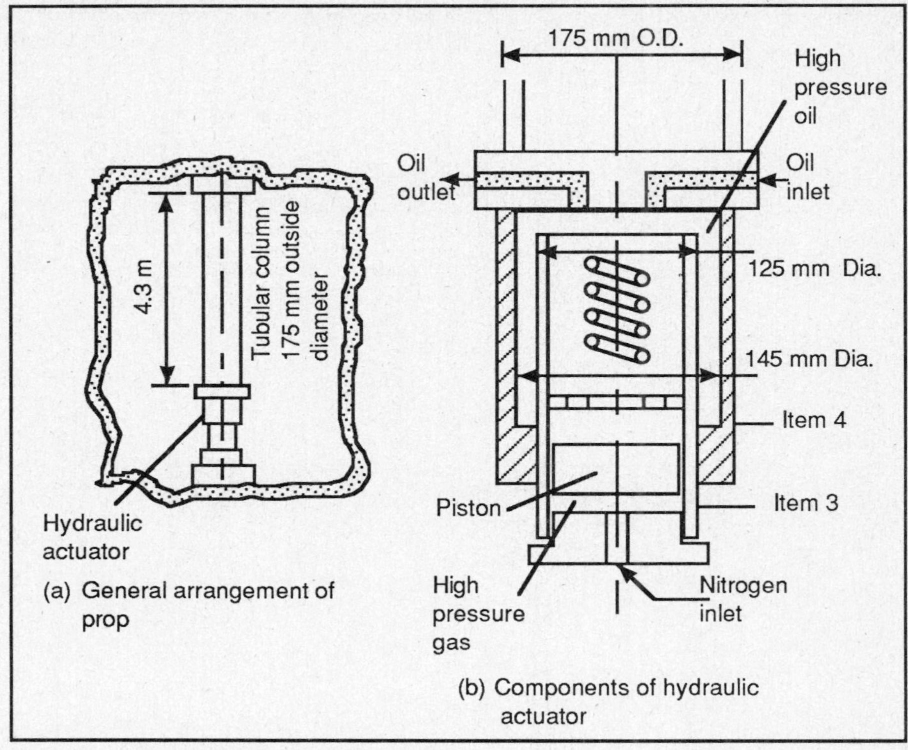

Figure 8.7 Schematic views of pit-prop

(a) investigate and determine how sensitive the allowable column load is to *dD* and *dt*; and
(b) for steel tubing of 175 mm outside diameter, given that $dD = 0.5$ mm and $dt = 0.1$ mm, revise your calculations, carried out in part 3 above, to take account of these tolerances.

8.7.6 Modular guard rail

Your company has decided to manufacture and market a modular guard rail/hand rail system. It has already been decided that the general configuration of each module is to be as shown on the diagram, Figure 8.8.

These modules are to be bolted (through the holes indicated) or welded to the side of existing decking. The top of the hand rail is to be approximately 1000 mm above the decking when installed. You may use material with the properties of ultimate tensile strength, $S_u = 350\ MPa$ and yield strength of $S_y = 240\ MPa$.

1. List possible and most probable modes of failure for the module. Determine suitable design factor of safety and determine the sizes of the uprights and cross straps.
2. Suggest ways of modifying the design to make more effective use of materials. Comment on the suitability, effectiveness and safety of the modules for a wide variety of uses, such as crowd control, playground or safety railing for small children and on piers in a marina for example. List other possible uses in similar vein.

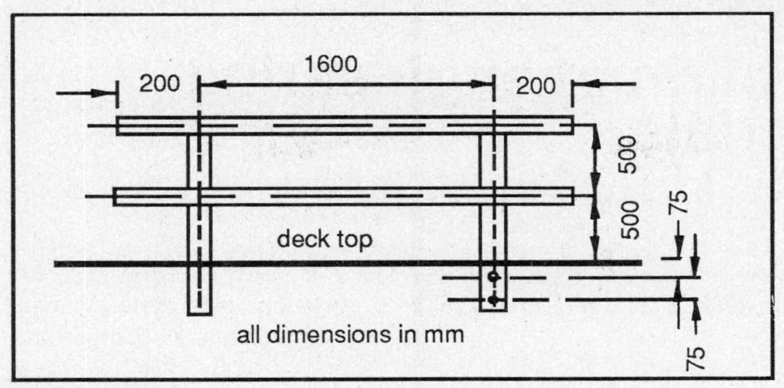

Figure 8.8 Modular guard rail
Note: All dimensions are in mm.

Notes: The designer has suggested three main uses for the railing module, namely:

(a) to prevent people from falling off elevated walkways and platforms;
(b) to provide hand support; and
(c) to instill confidence in people on elevated fixtures.

It has been decided to use material of rectangular cross section, and all modules will be treated to prevent corrosion.

You are not required to design the connections between members of the railing.

Chapter 9

Selection of engineering materials

...going in search of a great perhaps.
F. Rabelais 1553

Concepts introduced	selection of engineering materials based on material properties.
Application	spring material selection for an over-spin clutch.

9.1 Overview

Modern designers of products have access to a vast variety of materials and production processes. This is as it were the *'good news'*. The *'bad news'* is that the modern designer is expected to make a choice from materials and processes in a way which will allow for *'optimum performance'* of the final product. Optimum is undefined and in quotation marks since this property is invariably open to interpretation, Figure 9.1.

The ensuing pages set out a limited overview of materials, production processes and some limitations of both. It must be stressed that this text is not intended as a reference text but rather as an overview. All the information is aimed at providing the inexperienced designer with a *'ballpark'* feel for what is available. In all *'real'* design

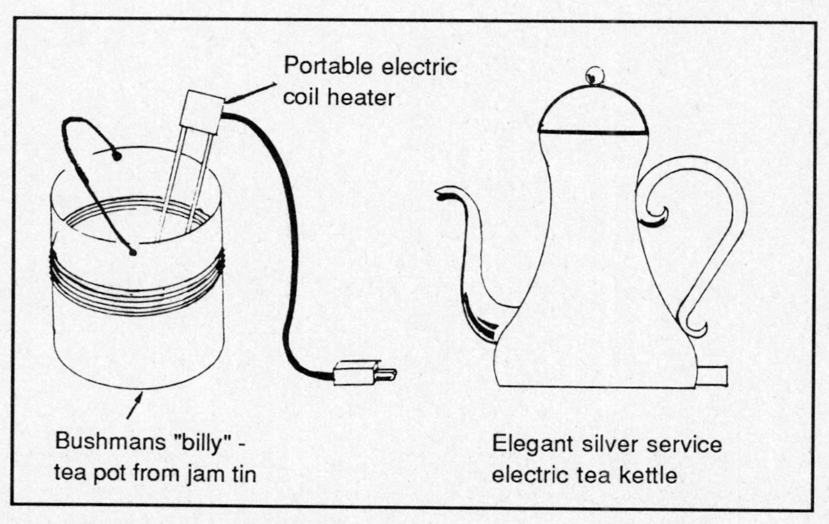

Figure 9.1 Two views of optimality

situations consultation must be sought with specialists in the relevant fields. No doubt in many cases there will be constraints on the problem that will preclude many of the available materials and processes. Nevertheless, the glimpse of the '*great perhaps*' may offer the designer a fresh new look at some old problems or old solutions.

9.2 Materials

Most engineering materials are classified by one of three designations, namely:

1. *Colloquial or trade name*. Examples are: *Teflon* for the product poly-tetra-fluor-ethylene (*PTFE*) and *Mild Steel* for a range of low carbon steels used in general engineering and designated by a typical 0.1 percent carbon content (for example: *A.S. 1442-1020*, or *AISI-SAE 1020*).
2. *Standard specification* according to either *Australian Standards* or some other internationally recognized standards such as *Society of Automotive Engineers* (*SAE*) or *American Iron and Steel Institute* (*AISI*).
3. *Trade specification* by companies producing special materials, usually prefaced by the company name. *Comsteel-444*(19/2), which is equivalent to *AISI-444*, ferritic stainless steel, is a good example of this type of designation.

Whatever the specification the designer should select a material that is both suitable and available. Consequently the process of material selection is based on consultation with the producer or distributor of the materials concerned. This will ensure up-to-date specification for the design.

Figure 9.2 charts the available alternatives in choosing engineering materials. Some select members of the set shown in Figure 9.2 will be dealt with in detail due to their special importance. Others will be neglected beyond an indication of the specialist areas of application.

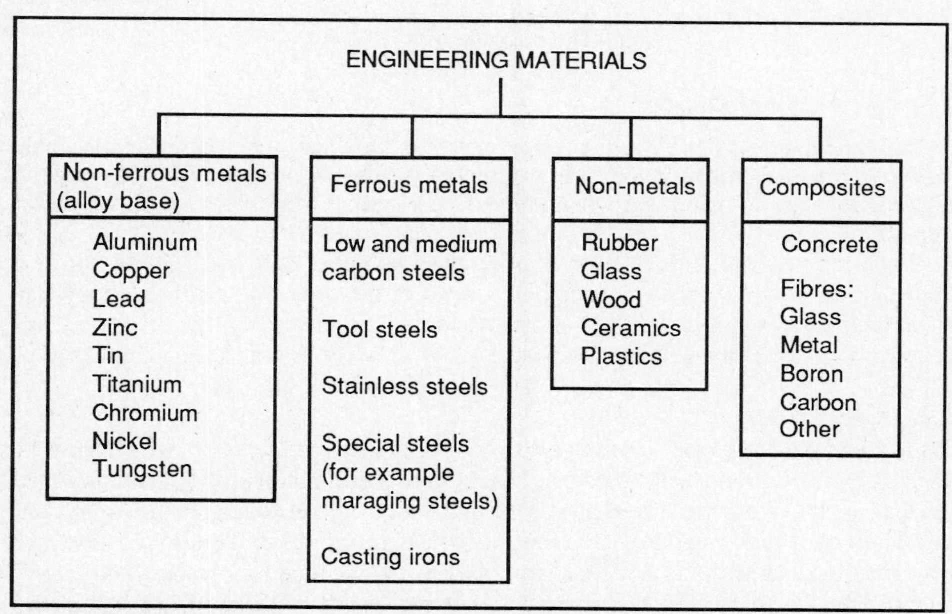

Figure 9.2 Common engineering materials

Materials are chosen for a variety of reasons but generally three dominant criteria are used in the selection process.

- First of all, the material *must meet the specified duty* for the product (strength, elasticity or wear resistance for example).
- Secondly, the material *must be readily available*. Material suppliers are notorious for listing exotic products which can only be produced if the buyer is prepared to wait many months or prepared to order a large tonnage.
- Thirdly, the material *must meet production requirements*. The product must be producible. It may be exciting to design a product in a special ceramic, but the excitement soon evaporates when it is discovered that the only furnace capable of producing the desired manufacturing conditions is in West Mongolia.

Table 9.1 gives the materials used in manufacture from 1944 to 1968 (US figures). As can be seen the rise of non-ferrous materials is steady but still, in 1968, a long way behind the total volume of grey cast iron or sheet and strip steel. It is also noticeable that of the non-ferrous metals aluminum is the most highly used. These figures simply underscore the general feel you may already have about the world of engineering materials. However, a cautionary note must be sounded about this. A number of years ago there was a general concensus among engineering designers of the 'old traditional school', expressed along the following lines:

> *When one designs some engineering device one calculates the loads and stresses on major components, then draws in the rest of the detail such that it **looks** right.*

This early, developed skill or concept must have filtered into the area of material selection also. Yet if one takes this idea and extrapolates back to its logical limit, we would all still be using stone axes and bows and arrows for hunting.

9.3 Iron and steel

Probably the most versatile of engineering materials, iron and steel have many desirable qualities. Iron is abundant in the form of iron ore (haematite or iron-oxide), it is readily reduced from ore to metal and when alloyed with carbon it provides a vast variety of properties.

When alloyed with carbon, iron is referred to as either *cast iron* ($C>3\%$) or as a *steel* ($C \leq 1.5\%$) — C denotes the carbon content of the alloy. Although steel may also be cast into shapes it is generally hot or cold formed or machined. Cast irons however are always cast into shape. (Note: The range $1.5\% < C < 3\%$ is not used commonly.)

9.3.1 Cast iron

Casting irons have a wide variety of excellent properties, some of which may be changed by appropriate heat treatment. Liquid iron is of a relatively high 'fluidity' and may be cast into quite intricate shapes. Due to the high carbon content in grey cast irons they have excellent wear resistance as well as high vibration damping. These two properties alone make this material most useful for machine tool frames and engine components. Due to uncertainty associated with the quality of the surface finish in cast components, unusually large machining allowances are called for when a component is

Table 9.1 Materials used in manufacture (Niebel and Draper, 1974, US data: 1ft^3 = 0.083 m^3)

Material	*1944* 10^6 x ft^3	%	*1948* 10^6 x ft^3	%	*1953* 10^6 x ft^3	%	*1958* 10^6 x ft^3	%	*1963* 10^6 x ft^3	%	*1968* 10^6 x ft^3	%
Ferrous												
Gray-iron cast	39.44		54.38		57.99		44.12		54.29		63.05	
Ferrous forgings	8.83		3.99		6.14		3.26		3.82		5.09	
Steel cast	7.84		7.49		7.78		4.76		6.39		7.36	
Semi-finished steel	31.97		16.2		18.9		10.31		13.42		20.50	
Steel shapes and plates	70.62		48.11		55.41		41.10		53.40		61.91	
Steel sheets and strip + other categories	51.9		81.47		112.8		94.21		131.9		138.6	
Total Ferrous	287.78	98.6	292.66	95.2	368.54	94.5	266.48	92.3	346.67	90.5	398.27	87.5
Non-ferrous												
Aluminium, cast and wrought	3.06	1.1	12.28	3.9	17.48	4.5	18.99	6.5	32.17	8.3	52.18	11.5
Magnesium	0.89	0.3	0.13	0.04	0.5	0.14	0.42	0.1	0.54	0.11	0.64	0.14
Total Non-ferrous	3.95	1.4	14.58	4.74	21.25	5.4	22.13	7.14	36.34	9.4	56.68	12.44
Plastics												
Phenolic					0.002		0.001		0.003		0.003	
PVC					0.001		0.002		0.003		0.007	
Styrene					0.003		0.002		0.01		0.02	
Polypropylene									0.002		0.005	
Polyethelene									0.008		0.03	
Total plastics					0.006	0.002	0.005	0.001	0.027	0.004	0.07	0.015

machined from a normal gravity type of casting. Machine finishing allowances for gravity cast irons is shown in Figure 9.3. Figure 9.4 shows schematically the production processes from iron ore to major commodities.

9.3.2 Plain carbon steel

Most engineering designers will recognize such terms as 'mild steel', 'tool steel' or 'spring steel'. However, the variety of steels commercially available for the production of commodities is vast and the range can be encompassed only through consultation with metals handbooks or manufacturers. The range is further complicated by the changes in properties available through heat treatment. This text is intended to provide only an overview of the range of steels. Consequently the brief survey that follows is broken up into the segments indicated on the 'tree of steels' shown in Figure 9.5.

Plain carbon steels range in carbon content from 0.1 percent to 1.5 percent with minor additions of manganese phosphorus, silicon and sulphur. To a very large extent the properties of the final product are affected by the way the steel poured from the furnace behaves during the cooling process. In the final stage of processing the molten steel is surrounded by a great deal of free oxygen. This oxygen reacts with the carbon in the steel and forms carbon monoxide (CO). As the steel cools the CO bubbles rise to the surface and the resulting metal structure is significantly less homogeneous than it would be without the bubbles. This material is known as rimming steel after the CO bubbling process (which is referred to as rimming). Rimming steels are suitable for low-strength cheap steel applications such as thin plates for metal cans. The thinning out process tends to reduce its performance in terms of structural integrity.

If the steelmaker wishes to produce a steel without the CO bubbles then additives are used at the last stages of production. Quantities of aluminum or silicon combine with free oxygen in preference to the formation of CO. The result is a very still or *killed* cooling process. In contrast to the rimming steel, fully *killed* steels tend to suffer cooling shrinkage. The lower volumetric efficiency of this production process results in a more expensive steel, but a steel vastly superior to rimming steel in structural properties.

The modern steelmaking process relies on continuous casting and in general additives are used to produce a fully killed steel. The terms *rimming* or *semi-killed* are only of relevance to the lesser-used ingot-making processes where the steel is cast in batches (*ingots*).

Practical uses of steels tend to fall into three major categories, namely *structural members*, components to be *heat treated* and *fabricated* components. For the structural components the property of weldability may be a major consideration in making the

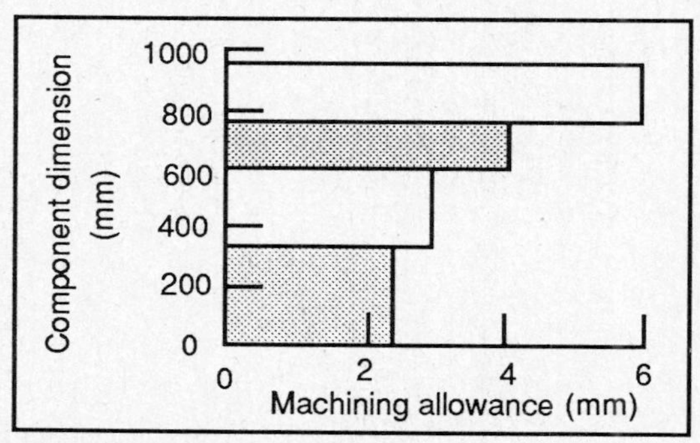

Figure 9.3 Machining allowances for cast iron

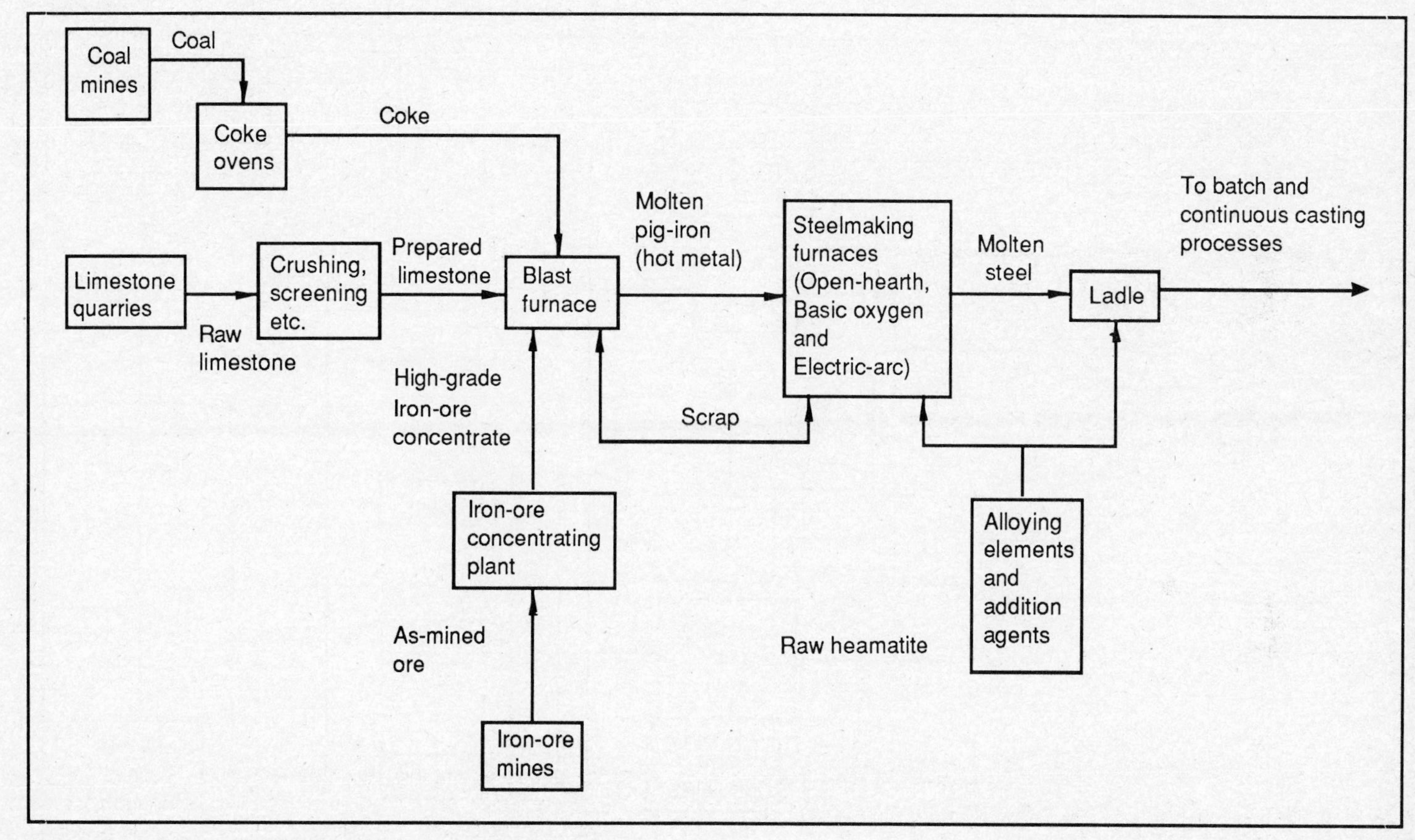

Figure 9.4 Ore to commodities (a) from ore to basic steel

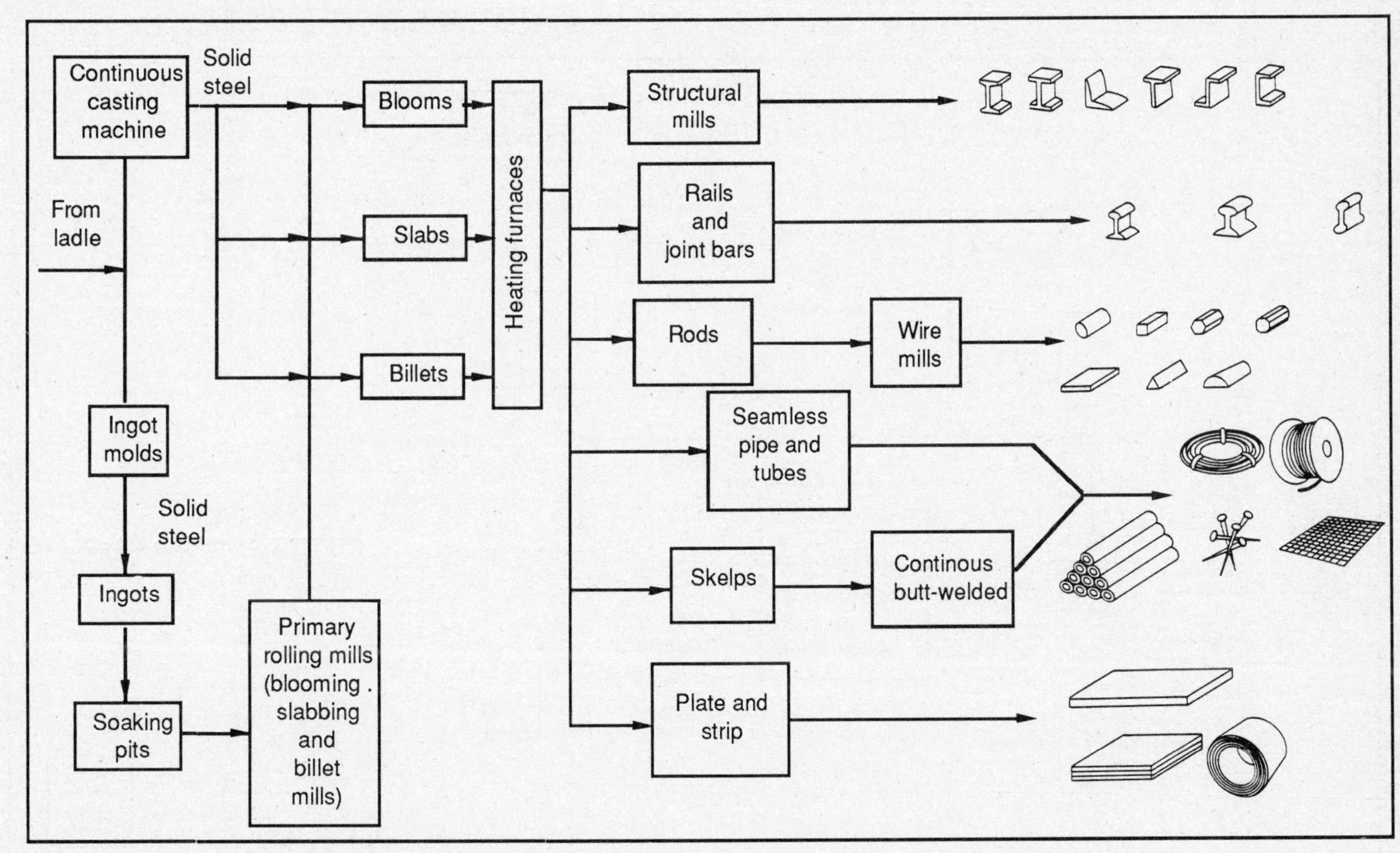

Figure 9.4 Ore to commodities (b) Milling of steel

proper material choice. Heat treated components on the other hand may require high strength and it is this property which will most influence material choice. Finally, for fabricated components the property of machinability may be of major importance. All of these properties may be enhanced by proper alloying of the steel in the production process.

The performance of steel may be enhanced in many other ways also by the addition of suitable base elements. One common additive however is sulphur in the range 0.15 percent to 0.35 percent to enhance machinability. The range of steels alloyed with this high sulphur content are referred to as *free-cutting* steels.

The heat treatment of steels is a special subject but in general terms the process is aimed at re-arranging the grain structure of the steel to enhance certain properties. Two major properties enhanced are hardness and toughness or lower susceptibility to shock load fracture.

The presence of carbon in steel is essential for the hardening process. In medium and

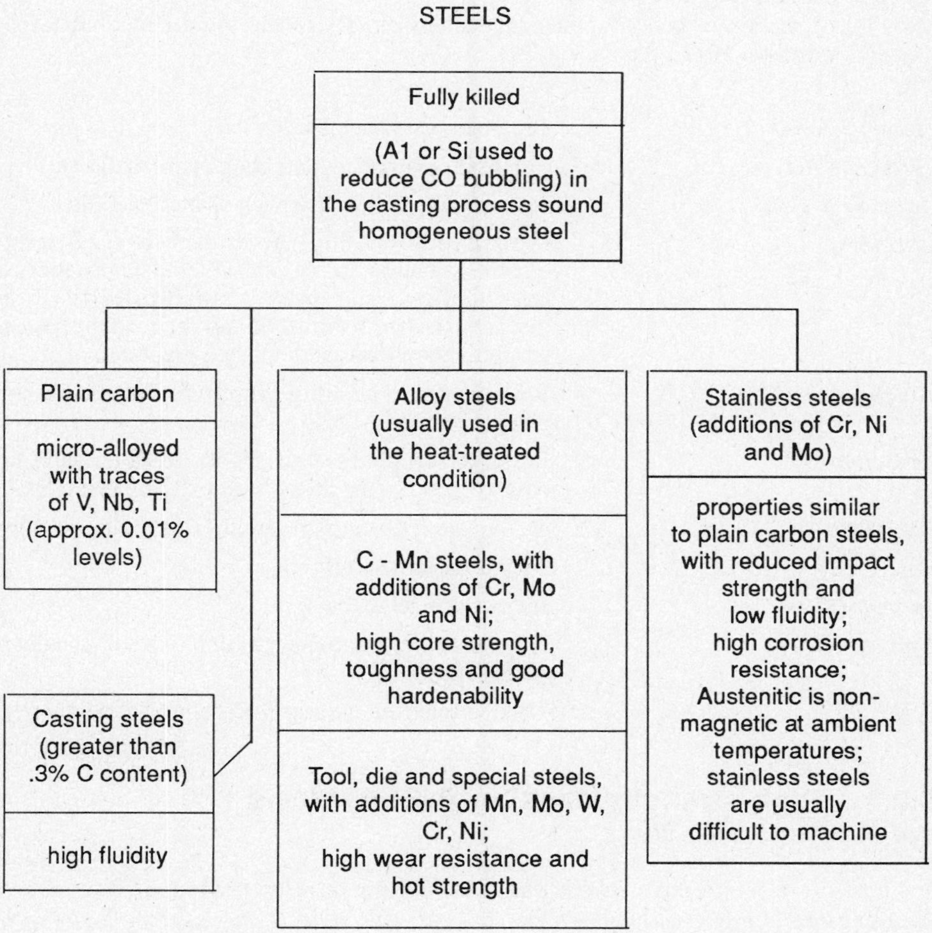

STEELS

Fully killed

(A1 or Si used to reduce CO bubbling) in the casting process sound homogeneous steel

Plain carbon

micro-alloyed with traces of V, Nb, Ti (approx. 0.01% levels)

Casting steels (greater than .3% C content)

high fluidity

Alloy steels (usually used in the heat-treated condition)

C - Mn steels, with additions of Cr, Mo and Ni; high core strength, toughness and good hardenability

Tool, die and special steels, with additions of Mn, Mo, W, Cr, Ni; high wear resistance and hot strength

Stainless steels (additions of Cr, Ni and Mo)

properties similar to plain carbon steels, with reduced impact strength and low fluidity; high corrosion resistance; Austenitic is non-magnetic at ambient temperatures; stainless steels are usually difficult to machine

Figure 9.5 The tree of steels

high carbon steels sufficient carbon is present for hardening to be achieved through a heat cycling process. This usually results in a very hard crystalline form of the iron-carbon solution *(bainite or martensite)*. For steels with a 0.2 percent or less carbon content the hardening thermal cycle is often preceded by a process of surface carbon absorption or *carburizing*. Surface hardening metals are usually referred to as having been *case-hardened*. Nevertheless, steels with such low carbon content can be through hardened.

Casting steels are similar in properties and composition to wrought steels, but the casting process reduces the impact strength of the steel. Furthermore the casting steels are less fluid than their cast iron counterparts. The resulting casting is rough with very large shrinkages. For these reasons casting tolerances and machining allowances must be correspondingly increased above those indicated in Figure 9.3 for cast irons.

9.3.3 Alloying steels

Alloying of metals in general enhances some property of the main constituent. The effects of principal alloying elements are:

Nickel (Ni)	Toughness and deep hardening.
Chromium (Cr)	Corrosion resistance, toughness, hardenability.
Manganese (Mn)	Deoxidation (killing), as well as hardenability.
Silicon (Si)	Deoxidation (killing), resistance to high temperature oxidation, raises critical temperature for heat treatment, increases susceptibility to decarburization (loss of surface carbon) and graphitization (formation of free graphite).
Molybdenum (Mo)	Promotes hardenability, improves hot tensile and creep strengths.
Vanadium (V)	Promotes toughness and shock resistance and hot strength; one of the three microalloying elements.
Aluminum (Al)	Deoxidation (killing), promotes fine grain structure;
Boron (B)	Increases hardenability;
Sulphur (S)	Increases machinability.
Tungsten (W)	Hardness / hot wear strength in tool steel; usually as tungsten carbide.
V, Nb, Ti	Used in trace amounts as microalloying elements.

9.3.4 High-strength low-alloy (HSLA) steels

Van Vlack (1985) notes that:

> *'Low-alloy steels have undergone some recent developments that have led to significantly higher strength products. ... structural steels that have yield strengths of greater than 500 MPa (> 70,000 psi) in contrast to earlier structural steels of only half that figure.*

These *HSLA* steels are alloyed with very small quantities (0.05 percent) of Vanadium, Niobium or Titanium to produce fine grain size and finely distributed precipitates within the ferrite. These precipitates reduce the formation of dislocations, and thereby provide the real strength of the alloy. The resulting steel may be used in the un-heat-treated state, making it attractive for use in the design of automobiles, high pressure pipe lines and major structures.

9.3.5 Stainless steel

This versatile high alloy material is best described in the words of Niebel and Draper (1974):

> *The term stainless steel denotes a large family of steels containing at least 11.5 percent chromium. They are not resistant to all corroding media... Stainless steel competes with non-ferrous alloys of copper and nickel on a corrosion-resistance and cost basis and with light metals such as aluminum and magnesium on the basis of cost and strength-weight ratio. Stainless steel has a number of alloy compositions and many suppliers. Information on its properties and fabrication can be obtained readily. Sound techniques have been evolved for casting, heat treating, forming, machining, welding, assembling, and finishing stainless steel. It will be found that this material usually work-hardens (which makes machining, forming and piercing more difficult), must be welded under controlled conditions, and under inert gas. It has desirable high strength, corrosion resistance and decorative properties.*

Stainless steels are hardenable by cold working and the properties of the steel may be enhanced through appropriate heat treatment, cold working or a combination of these techniques. The single most important characteristic of stainless steels is their resistance to corrosion. This resistance is due to a thin transparent chromium oxide film which forms on the surface of the metal on exposure to oxygen or air. Although protection is provided against most oxidizing substances some further surface protection is necessary when stainless steel is used with reducing agents such as hydrochloric acid. In situations where both hot strength and low surface oxidation rates are needed, a stainless steel with a ceramic coating may be used.

In some applications stainless steels are used for their appearance and aesthetic qualities rather than their corrosive resistance. The surface will accept a high polish and some steels may be readily moulded by mechanical deforming processes.

9.3.6 Special steels

Tool and die steels are a group of materials developed for the forming and machining of metals. Hardness, toughness and high hot strength are the major characteristics of these steels. Carbon content varies from 0.25 percent to 2.35 percent and chromium, vanadium and tungsten are the main additives to enhance material properties as described earlier. Typical trade designations for such steels are very much descriptive of their intended application:

- *High speed steels*: For general purpose use in drills, cutting tools and highly abrasive cutting conditions.
- *Hot work steels*: Used for Al and Mg extrusions, press forging dies, shear blades for heavy work, plastic moulding dies, forging dies, high termperature and abrasive conditions, extrusion tooling and copper and brass die casting.

- *Cold work tool steels*: General purpose tools, air and oil hardening tools, punches, dies, wear resistant tools such as brick moulds, crusher rolls and blanking tools.
- *Shock resistant steels*: Shock resistance is a prime property of these tough steels. They are used in pneumatic chisels, shear blades, hand and machine punches, chisels and dies.
- *Carbon tool steels*: Less expensive low alloy high carbon steels used where service conditions are not severe, for example in scale pivots, gauges and precision engineers' tools.
- *Mould steels*: Medium carbon chromium alloy steel for high pressure zinc die casting, plastic moulds and valve spindles. These steels have good corrosion and abrasion resistance.

9.4 Non-ferrous metals

The three most common non-ferrous metals alloyed to form metal products are aluminum, copper and nickel.

9.4.1 Aluminum

The alloys of aluminum are particularly useful due to their corrosion resistance, malleability and fluidity as casting materials. High strength to weight ratios available with aluminum make these alloys particularly useful in the aircraft aero-space industry. Hardening of the alloy may be achieved by either heat treatment followed by aging or by cold working or strain hardening. Aluminum may be *cast*, *rolled*, *extruded*, *forged*, *spun* or *shot-peen deformed*. Apart from steels aluminum is the single most useful structural metal.

9.4.2 Copper

Copper is available as a commercially pure base metal for conductors or alloyed with zinc (*brasses*), tin, aluminum and silicon (*bronzes*). Brasses are used for moulding and deep drawing. Although some work hardening takes place during cold working, the metal is readily annealed at moderate temperatures and successive stages of pressing can result in very large aspect ratio (depth to diameter ratio) pressings. Typical examples are cartridge cases and cosmetic containers.

Brass type *360A* (American Society for Testing of Materials designation) is the most common type of brass available and it is referred to as free machining brass. This alloy is one of a series of '*leaded-brasses*' containing about 2 percent by weight of lead. This alloy is probably the most easily machined metal known to man and it is generally used as a basis for machinability comparison.

Bronzes are most useful materials where corrosion resistance is important. Main uses are valves, naval applications, propellers and general purpose bearings. Phosphor bronze is a particularly hard alloy and in general its use is as corrosion-resistant spring material. Due to its resistance to sea water corrosion and its high strength it is also useful as a casting alloy for high-power marine propellers. Nickel bronzes are used mainly in food machinery.

9.4.3 Nickel

The most common types of nickel alloys are *Monel 400* (66.5/31.5 percent copper-nickel alloy with nickel the main constituent), *Incoloy 825* (38 – 45 percent nickel and the remainder iron) and *Inconel 600* (72 percent nickel with 15 percent Cr and 8 percent Fe). Table 9.2 lists some properties for these alloys. These alloys are particularly useful for their corrosion and heat resistant properties. Common uses are:

- *Monel*: Protective plates in chemical, marine and steam plants. Good corrosion resistance to sea water, sulphuric, phosphoric and hydrochloric acids as well as fatty acids.
- *Incoloy 825*: Heating element sheaths, process piping, heat treatment containers.
- *Inconel 600*: Furnace parts, heat treatment equipment, electrical heater elements.

Table 9.2 Most common nickel alloys

Metal	*Typical Composition (%)*							
	Ni	C	Cu	Cr	Fe	Rem	S_u MPa	S_y MPa
Monel 400 ASTM B 127 B163-5	66.5	0.2	31.5		1.2	Mn,Si	483	193
Incoloy 825 ASTM B163 B407-9	32.5	0.05	0.4	21.0	46.0	Al, Mn Si, Ti	586	241
Inconel 600 ASTM B163 B166-168	76.0	0.08	0.2	15.5	8.0	Mn, Si	552	242

Key: S_u = Ultimate tensile strength;
S_y = Yield strength.

9.5 Plastics and other composites

Table 9.3 provides comparisons of some mechanical properties of various reinforcing fibers and composites.

By contrast to metals, plastics are weak and compliant. The most difficult feature of designing with polymers is the time-dependent nature of their structural properties, such as yield strength for example. Coupled with these features, plastics are durable, often corrosion-resistant and in general electrically insulating. In special cases they can have superior structural properties to metals (fatigue life of polypropylene for example).

The early use of plastics was due specifically to their easy low temperature mouldability and relatively low cost for mass-produced components. It is, however, in the composite fiber reinforced form that plastics have come of age as a significant structural material. The earliest reinforcing used was glass (glass in polyester or epoxy resin for fiber-glass), but as the technology of fibers advanced, metals and carbon reinforced composites became the most advanced structural materials. This development reflects the constant striving of designers for lighter and stronger structures in the automobile, aircraft, space craft and recreational sporting industries. The most desirable property of these composites is the ability to engineer with them almost endless combinations of directional strength and stiffness behaviour.

Table 9.3 Mechanical properties of fibers and composites

Fiber	*Tensile strength MPa(ksi)*	*Tensile modulus GPa(10^3 ksi)*	*Density (tonnes/m^3)*
Glass	1500(218) – 4300(624)	70(10) – 86(12.5)	2.5
Carbon	2000 (290) – 3500 (508)	200 (29) – 400(58)	1.7 – 2.02
Boron	35000 (508)	420 (61)	2.65
Kevlar	2700 (392)	60(9) – 130(19)	1.45
Polyester matrix	20(2.9) – 40(5.8)	1(0.15) – 3(0.45)	1.4 – 2.2
Epoxy matrix	40(5.8) – 90(13)	1(0.15) – 4(0.6)	1.6 – 1.9
High carbon steel	2800(406)	210(30)	7.8
GFRP(cloth, 50%)	240(35)	14(2)	1.7
GFRP(unidirectional – 50% glass)	1200(174)	50(7)	2.0
Carbon-fiber R.P. – (unidirectional HT fiber)	1600(232)	129(19)	1.5

Source: Crane and Charles, 1984.

9.6 Wood and concrete

Wood was probably the first complex structural material used by *Homo Sapiens*. Due to its nature timber is anisotropic and quite variable in structural properties even within one species. The most common mode of failure in timber under compression is the buckling of wood fiber (held in a resin matrix). For this reason timber is much stronger in tension than in compression. Standardised structural grading allows for predictable performance in timber structures. Laminated, either as beams or plywood, timber properties can be significantly enhanced over the bulk properties. In general structural timber today is used mainly in the housing industry for roof structures, most commonly as ‘gang nailed’ trusses or as laminated beams.

Concrete is a mixture of four components, namely sand, cement, stone and water. The main attraction for its use lies in the ability to flow it into shape . Concrete develops its strength over a period of time after pouring (up to 60 MPa after 28 days in compression). Due to its brittle nature, concrete is weak in tension and in structural use the tensile loads are taken by combining the concrete with reinforcing materials, usually structural grade steel.

9.7 Material selection

Selecting the right material for a specific design is probably the single most important decision facing designers. Unless other constraining factors pave the way to a particular choice the following recipe will generally result in a short list of hopefuls from the ocean of materials:

1. *Will it work*? The answer to this question rests with the matching of dominant or primary criteria such as strength, hardness, elastic behaviour (Young's or Shear moduli), hardenability, toughness, magnetic properties, thermal and electrical conductivity.
2. *Will it last*? This question is answered by the important criteria of corrosion and heat resistance as well as resistance to wear, dynamic loading, shock, creep and stress corrosion.
3. *Can it be made*? This question refers to castability, machinability and surface finish performance or surface enhancement such as coating, plating and anodizing.
4. *Can it be done within specified limits*? The main constraints of most engineering components or sructures are cost and weight. For these constraints to be met, the designer needs a feel for relative costs and relative weights.

The answer offered here to question 4 is rather a simple view of the most common structural requirements. In the most general case the limits on material performance will include many factors apart from weight and cost. Crane and Charles (1984) attempt to formalize the material selection procedure, but they too present a table of parameters based on cost and weight for a great variety of simple engineering components. Table 9.4 is based on this approach to material selection, listing simple structural components and the critical parameters to be used in a minimum mass design, depending on whether the component is stress or deflection limited.

Figure 9.6 shows the case of a simple rod in tension under the action of a load P. For this case, the minimum mass design is evaluated as follows:

$$\text{mass of rod} = \rho \times l \times A$$

1. stress limited design:
minimum mass (per unit load and per unit length) $= \rho_{min} \times A_{min} = \dfrac{\rho_{min}}{(S_y)_{max}}$

2. deflection limited design:
minimum mass (per unit load, length and deflection) $= \rho_{min} \times A_{min} = \dfrac{\rho_{min}}{E_{max}}$

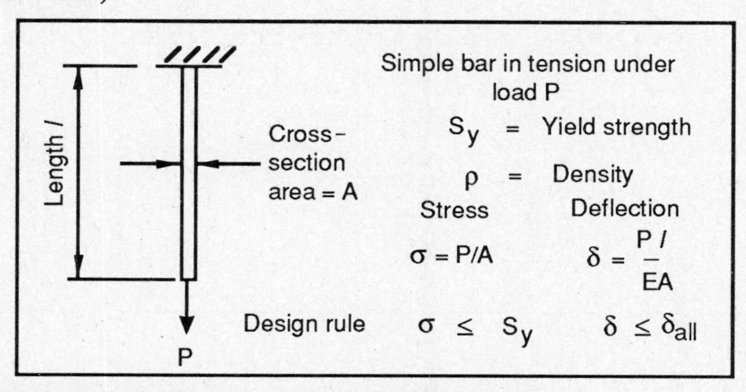

Figure 9. 6 Design parameters for a simple rod in tension

Table 9.5 Metal properties

Metal	*Dominant properties and uses*	*relative cost*	*Mechanical and physical properties*									
Steels			E GPa	G GPa	S_u MPa	S_y MPa	S_e MPa	H_B	μ	δ %	SG	α $\times \frac{10^6}{°C}$
S1 AS1422-CS1020, commercial , structural grade, semi-killed	Hot-rolled sections, bars, plates	1	208	81	380	240	$.5S_u$	126 179	.27	22	7.84	3.61
S2 AS1442-CS1042, mild steel	General fabrication bars and plates	1.04	208	81	540	270	$.5S_u$	152 207	.27	16	7.84	3.61
S3 AS1442-CS4140, C - Mn, fully killed	General purpose, highly stressed structural components, bars, plates	1.1	208	81	540	300	$.5S_u$	269 331	.27	18	7.84	3.61
S4 COMSTEEL 4140, SAE 4140, BS970/708-M40	As for S3, but supplied hard, tempered and/or ground to H11	4.5 5.7	210	81	700 1080	525 755	$.5S_u$	201 331	.27	12 17	7.80	4.61
S5 Stainless AISI 304, austenitic, strain harden only	Good deformability and surface-finish, process equipment, brewers vats	8.42	196	87	520 1280	205 965	$.35S_u$	183 330		8 40	7.80	5.34
S6 Stainless AISI 316, austenitic	High-temperature corrosion resistance, chemical lab. equipment	13.56	196	87	520	205		217		40		
S7 Stainless AISI 431, martensitic, harden by heat treatment	Excellent hardening properties, pump shafts, valves	6.20	196	87	520	205		183		20		

S8 Stainless AISI 440, martensitic	High carbon content high final hardness, for bearings, valve parts	5.97	196	87	520	205		183		20		
S9 Grey cast-iron	High fluidity, gravity castings, engine parts	0.8	96 131	60.7	140 415		69 165		.25		6.95 7.48	3.33
S10 Nodular cast-iron	Similar to S9, but better fatigue properties	1.1	162	60.7	400 700				.25		7.12	3.67
Non-ferrous alloys												
A1 Aluminum 2011-T3 (T3 refers to temperature properties)	Free machining, good corrosion resistance, general machined products	9.56	73	27.6	275	124	117	40	.3	14	2.7	7.22
A2 Aluminum 6063-T5	Extrudes well, weldable anodizes well	8.46	73	27.6	151	110	75	25	.3	8		
A3 Aluminum 6351-T5	Good corrosion resistance, welds well, medium strength, structural	7.96	73	27.6	262	241	130	40	.3	8		
A4 AS1565-801A Tough-pitch copper	Commercially pure casting and extruding metal, high conductivity switch gear and wire	7.53	131	48.3	210 390	70 350		40		8 20	8.23	9.17
A5 AS1565-851D Brass	Free machining,valves, general castings	6.69	103	38	390	170	$.35S_u$	45		12	8.42	6.55
A6 Magnesium usually alloyed with aluminum (3-10%)	Light, easily machined and formed	12	45	16.5	380	260	110				1.8	8.89
A7 Titanium alloy(5%Al, 2.5%Sn)	Light, rigid metal, aerospace applications	26	117	42	800	750	$.6S_u$		.33		4.46	3.16

Key:

E, G = Young's and shear moduli
μ = Poisson's ratio
HB = Brinnel hardness
δ = elongation at failure
SG = specific gravity
α = the coefficient of thermal expansion

Table 9.5 is a tabulation of various material properties. Wherever possible the materials listed are similar to those specified by the *Australian Institute of Metals and Materials Handbook (USA: ASM-International Handbook)*. Strength properties are all approximate values and final selection must be made by reference to either the maker/supplier specifications or to independent test results. In this context the designer may specify a material either by analysis (if future heat treatment is intended for the component) or by minimum strength properties (if the component is to be used in the 'as supplied' form).

Table 9.4 Minimum mass criteria for different loading conditions (Crane and Charles, 1984)

Component	*Stress limited*	*Deflection limited*
Rods in tension	ρ/S_y	ρ/E
Thin-wall pipes and pressure vessels under internal pressure	ρ/S_y	
Flywheels for maximum kinetic energy storage at given speed	ρ/S_y	
Helical spring for specified length and load capacity	ρ/τ_{all}	
Shaft in pure torsion	ρ/τ_{all}	ρ/G
Beam with fixed geometry in bending	$\frac{\rho}{(S_y)^{2/3}}$	$\frac{\rho}{\sqrt{E}}$

Key: τ_{all} = allowable shear stress;
G = shear modulus.

9.8 Notes from a designer's workbook

The following problem is a case study in yield limited design in which you are required to select the most appropriate material.

9.8.1 Design of a leaf spring for an overspin clutch

A leaf spring is usually designed to operate as a simply supported beam with a normal load applied at mid span. It is an essential feature of such springs that they are designed to operate fully within their elastic range. Any 'permanent set' of the spring will render the operating behaviour (assumed to be linear) useless.

Consider the design of a clutch (friction plate between prime mover and load) so that when the angular velocity of the prime mover exceeds some pre-determined value the clutch 'releases'. With such a design the load can never exceed the preset angular velocity. Such a device is called an 'overspin release clutch', see Figures 9.7 and 9.8.

You may assume that the springs holding the clutch plate engaged during normal operating speeds are rectangular section leaf springs of length *L* mm between simple supports. The motor driving the load is 15 kW operating at 800 rpm normally. The clutch must fully release at 1000 rpm. Select the 'best' material for the cluch spring.

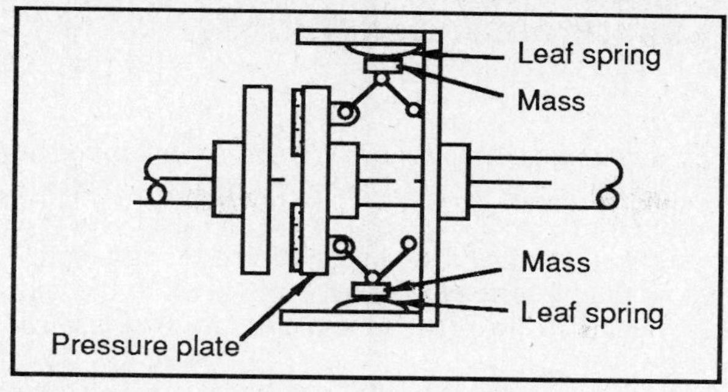

Figure 9. 7 Schematic view of overspin release clutch

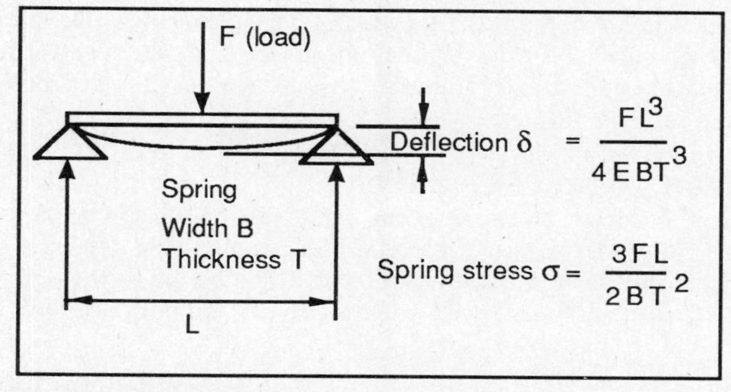

Figure 9. 8 Schematic view of spring

The design rule for the spring is that the yield strength should not be exceeded and the design stiffness (δ/F) should be met. Substituting from the deflection equation for δ/F into the stress equation we get:

$$\frac{S_y}{E} \geq \frac{6\delta T}{L^2}$$

Now we can tabulate the left-hand side of this inequality for a variety of materials and find the 'best' for this application.

Table 9.6 Properties of spring materials

Material	S_y *MPa*	*E* *GPa*	$\frac{S_y}{E}$ ($\times 10^{-3}$)
Brass (cold-rolled)	638	120	5.32
Phosphor bronze	770	120	6.43
Beryllium copper	1380	120	11.5
Spring steel	1300	210	6.19
Stainless steel	1000	200	5.0
Nimonic alloy (high-temperature)	614	200	3.08

9.9 Exercises in material selection

9.9.1 Flywheel

In a high efficiency experimental road vehicle some of the drive energy is stored in a flywheel. The flywheel is designed to operate at 20,000 rpm. The flywheel is designed to be a ring shrink-fitted over radial spokes. When the speed exceeds 21,000 rpm, the ring is free to slip on the spokes and is no longer driven. Assuming that the major part of the deflection from minimum to release speed takes place in the spokes and the spoke stress should always remain in the elastic range, select the best material for the spokes.

9.9.2 Pressure vessels

An aircraft fuselage, a scuba diver's oxygen tank and road tanker carrying liquid petroleum gases are all examples of pressure vessel designs. Develop criteria for choosing the 'best' material for the applications noted. Compare results for these vessels given both minimum weight and minimum cost as two possible alternative criteria for selection.

9.9.3 Ocean going LPG tanker

You are required to design an ocean-going liquid petroleum gas tanker. Explore (at a conceptual level) the best possible material choices available for such ships. In your evaluation consider the various functions to be fulfilled by the ship, namely containment (C), insulation (I), structural resistance to pressure (S) and normal functions of the ship hull (H).[1]

9.9.4 SCUBA tank

Design of a tank for self-contained under-water breathing apparatus (SCUBA).
Your task is to evaluate the parameters of design for the air tank carried strapped to the back of the diver.

It has already been decided that the design depth for the SCUBA gear is to be 200 meters and that the contained volume of the tank is to be 12.5 liters. The tank will be filled on land so that the internal pressure of compressed air in the tank will always exceed the local pressure of sea water when diving.

The design is only at the preliminary stage and you may make the following assumptions:

(a) The tank may be regarded as a plain cylindrical pressure vessel of diameter D with hemispherical ends and the design may be based on the failure of the cylindrical section of the tank.
(b) The tank wall thickness (t) will be uniform everywhere and relatively thin compared to the tank diameter.
(c) The volume of material in the tank may be approximated as the product of internal surface area times the wall thickness. The relevant geometric approximations are (for $D \gg t$):

$$\text{volume of hollow cylinder} = \pi D L.t$$

$$\text{volume of hollow sphere} = \pi D^2.t$$

[1] The authors are indebted to Professor Michael French for this problem, based on the presentation in M. J. French, *Conceptual Design for Engineers*, the Design Council, London, 1985.

Develop an expression for the empty weight, W, assuming that the shell thickness is equal to its minimum allowable value. Your expression should be in terms of the following symbols:

ρ_m = material density.
g = gravitational field constant.
p = design pressure for tank.
f = working design stress for the shell material (S_y/F_d)
η = shell joint efficiency.
C = total (inside and outside) corrosion allowance.
V = contained volume of tank.
D = internal diameter of tank (independent design variable).

Note that the length, L, of the cylindrical part of the vessel and the shell thickness, t, will need to be eliminated somehow from the expression you develop.

Determine the optimum proportions of the tank, when the corrosion allowance is negligibly small (compared to 't'). Optimum here means a tank of minimum weight. State all the design constraints that need to be imposed to achieve realistic values for the tank's proportions.

Select the most appropriate material for the tank, recognizing that the weight is of utmost importance for the user. Using the data from Table 9.5 compare the performances of stainless steel, aluminum alloy and titanium alloy. Naturally you must take material costs into consideration.

Chapter 10

Fatigue failure and stress concentrations

God is subtle, but not malicious.
Albert Einstein

Concepts introduced	dynamic loading; mean and alternating components of stress, fatigue failure endurance limit; surface finish and geometric stress factors.
Methods presented	safe life and fail safe design; A-M diagram; modified Goodman diagram.
Application	design of a jack-hammer.

10.1 Dynamic loading: Fatigue

10.1.1 Introduction

Conditions of fatigue loading

Engineering elements under dynamic loading may fail by fatigue. The concept of *fatigue* is used to denote the deterioration in strength of an element as a result of the repeated application and removal of load, so that it fails with little or no warning under a load which it could support indefinitely if statically applied.

Fatigue loads are classified according to the way in which they vary with time. Important types of fatigue loading are:

1. Regular cyclic stressing (illustrated in Figures 10.1 and 10.2):
   (a) sinusoidal — constant amplitude:
      (i) fluctuating stress;
      (ii) repeated stress;
      (iii) reversed stress.
   (b) sinusoidal — varying amplitude;
   (c) non-sinusoidal.
2. Irregular or random stressing.

(Fatigue under randomly varying stress is a specialized subject in its own right, outside the scope of this book.)

Commonly used symbols are:

σ_a = alternating component of stress or stress amplitude.
σ_m = mean stress.
σ_{max} = maximum stress during cycle.
σ_{min} = minimum stress during cycle.

$$R = \frac{\sigma_{min}}{\sigma_{max}} = \frac{\sigma_m - \sigma_a}{\sigma_m + \sigma_a} = \text{stress ratio.}$$

Most of the discussion which follows will be devoted to applications where the fatigue life is high, usually greater than 10^5 cycles. This is referred to as *high-cycle* fatigue. However, in some applications such as pressure vessels and associated pipework and turbine rotors, a structure or machine may be subject to a comparatively small number of load cycles during its working life. (For example, a pressure vessel used in a manufacturing process involving two cycles per day would experience less than 25,000 cycles in 30 years.) It then becomes necessary to consider the problem of *low-cycle* fatigue, another specialized subject (Faupel and Fisher, 1980; Langer, 1971).

The analytical methods to be presented here support experimental methods but do not replace them. They enable the designer to make an early assessment of a new design to eliminate weaknesses and highlight critical sections where tests may be called for. If one or more alternative designs are to be evaluated, those with low calculated fatigue lives can be discarded. Further, if an existing design is to be improved perhaps to remedy a premature failure, then analytical methods enable the relative effectiveness of different modifications to be estimated.

Design philosophy

Machine elements are almost always designed on a 'safe-life' basis, but structures may be designed on either a 'safe-life' or a 'fail-safe' basis.

Safe-life design relies on the ability to predict the fatigue life of the element or structure, so that before the end of this time the element or structure can be repaired, replaced or retired. The designer must have sufficient confidence in the method of prediction in order for the probability of a fatigue failure occurring to be extremely remote.

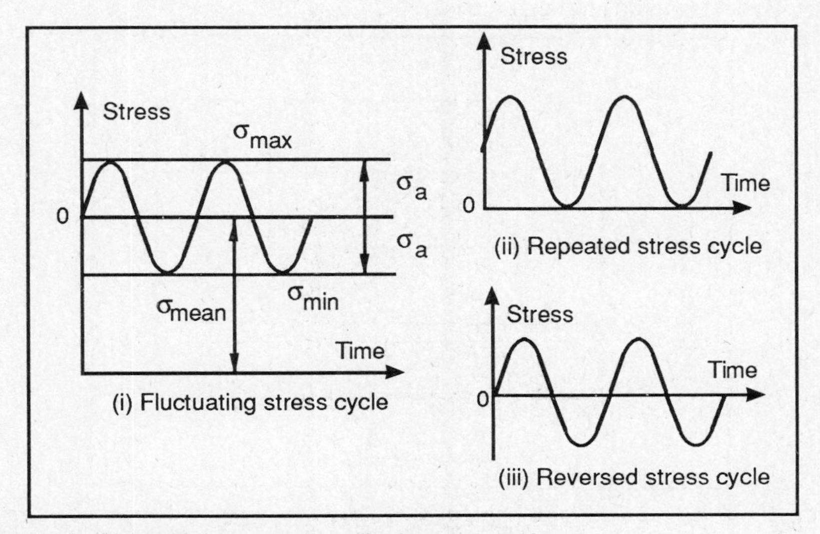

Figure 10.1 Sinusoidal constant amplitude stress cycles

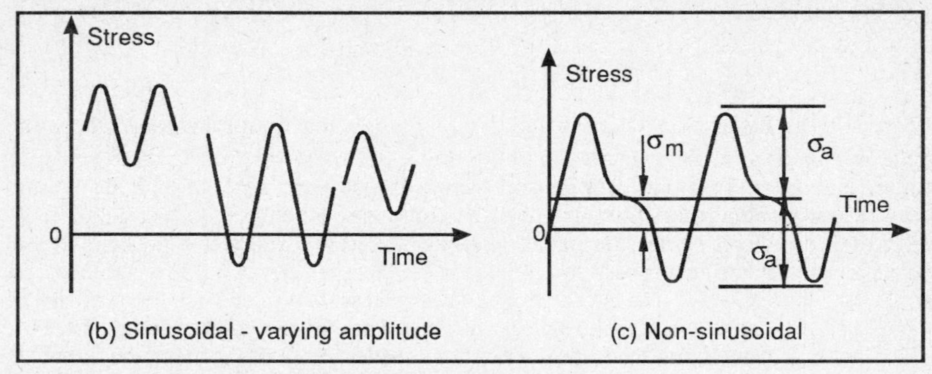

Figure 10.2 Other forms of regular stress cycles

alternate load-carrying members is sufficiently high and the rate of crack propagation sufficiently low for the structure to continue in service until such time as the cracks are discovered at a scheduled inspection. Features of the design are that the structure is redundant, that it is subject to regular inspection and that any fatigue cracks which do form are readily detectable before they become catastrophic.

Experimental data

To determine the properties of a material under fatigue loading, carefully prepared specimens are subjected to sinusoidally varying stress and the number of cycles to failure measured for a range of stress amplitudes (BS 3518). Basic fatigue data are obtained under conditions of reversed, uniaxial stressing (zero mean stress) and presented in the form of *S/N diagrams*. The number of cycles to failure *N* is plotted on a logarithmic scale on the horizontal axis and the stress amplitude S on the vertical axis. The scale for S may be linear or logarithmic, a logarithmic scale being preferred if it enables the results to fall on a straight line.

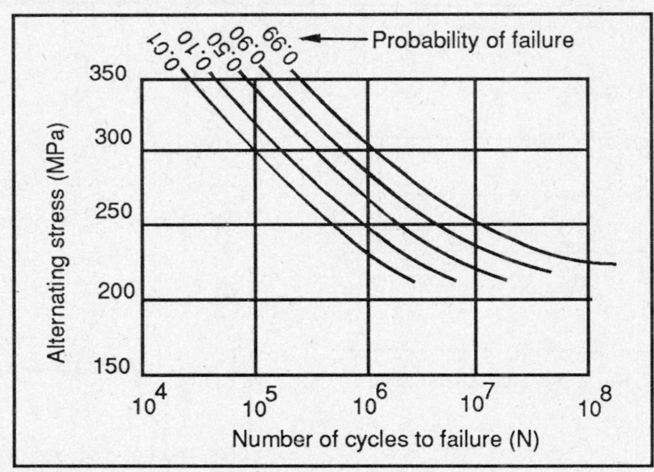

Figure 10.3 P/S/N curves for extruded 2024 Al. alloy unnotched specimens, rotating beam test

Source: Adapted from Mann, 1967.

Most steels exhibit a *'fatigue limit'* or *'endurance limit'* and if the amplitude of the reversed stress cycle is below this limit no failure will occur. Most non-ferrous alloys do not have an endurance limit and it is then common practice to define their fatigue strength by the stress level to cause failure after 10^8 cycles of load reversal.

The results of fatigue tests show pronounced scatter because fatigue performance is sensitive to very small variations in specimen preparation and metallurgical structure and in testing technique. If enough tests are carried out, statistical analyses can be made and the results expressed in the form of a *P/S/N diagram*, as shown in Figure 10.3.

Figure 10.3 indicates the probability of specimen or component achieving a specified endurance, and the prediction of the safe-life of an element of a structure or machine can then be related to an acceptable probability of failure.

In fatigue tests under fluctuating stress, the mean stress is not zero; experimental data can be conveniently shown on an *A/M diagram* with alternating stress on the vertical axis and mean stress on the horizontal axis, as shown in Figure 10.4.

We now examine methods available for designing to resist fatigue. The discussion will be limited to cases of uniaxial stressing. (For consideration of combined stresses, see Shigley, 1986, Sections 7-16).

10.1.2 Design of elements subject to reversed axial stress

The endurance limit S'_e (or fatigue strength at a specified number of cycles to failure) of a carefully prepared specimen of ferrous material is known, and we now wish to relate this to the fatigue strength of an element in practice, S_e.

The fatigue strength of a component in a practical application is affected by a large number of factors:

1. *surface finish*;
2. *presence of stress concentrations or notches*;
3. *size*;
4. *environment*:
   (a) temperature;
   (b) corrosion;
5. *process of manufacture*:
   (a) residual stress;
   (b) metallurgical condition of surface layers of material;
   (c) surface hardening process;
   (d) protective surface coating;
6. *metallurgical factors*:
   (a) inclusions and porosity;
   (b) grain size;
   (c) directional effects, anisotropy;
7. *mechanical factors*:
   (a) pre-loading;
   (b) shrink fits, press fits;
   (c) frequency of load cycle.

In predicting the fatigue life of a component we must also consider:

8. *scatter in experimental data*.

We now review evidence on four important factors from the above list, namely items 1, 2, 3 and 8. It is possible to make some useful generalizations concerning these factors.

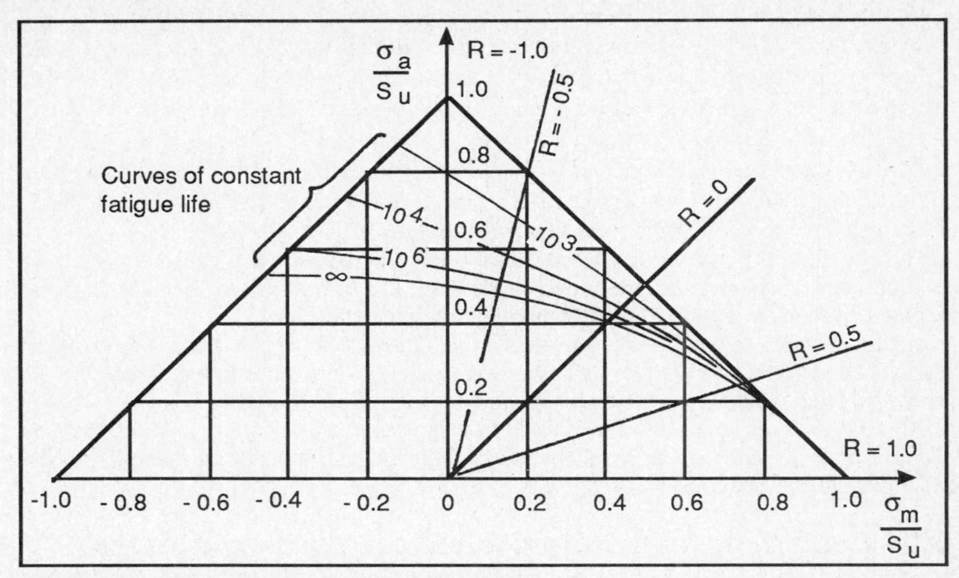

Figure 10.4 A/M Diagram for steels, unnotched
Source: Adapted from Heywood, 1962.

There is a great deal of specialized experimental data about the other factors, and the reader is referred to Faupel and Fisher (1980), Frost et al. (1974) and Osgood (1970) for helpful discussion and detailed information.

Surface finish

Most fatigue failures are initiated from surface defects, and the better the surface of a component, the higher its fatigue strength may be expected to be. However, experimental data do not only depend on surface finish but also on the nature of the residual stresses remaining in the surface layers of material as a result of the manufacturing process employed. B.S. 3518 specifies that the surface roughness of specimens is not to exceed 0.12×10^{-6} meters (center-line average) and prescribes the sequence of machining and polishing operations to produce this finish. For a particular type of manufacturing process the fatigue resistance of a material usually decreases as the profile of its surface becomes rougher; see Frost et al. (1974) and Shigley (1986). Data from these authors enables us to make an estimate of K_s, the fatigue strength reduction factor for surface finish.

$$K_S = \frac{\text{endurance limit of laboratory specimen with smooth surface}}{\text{endurance limit of specimen with rough surface}}$$

Stress concentrations due to changes in geometry

Stress concentrations arising from changes in the geometry of a part take three main forms: shoulders, holes and vee-grooves. They are often referred to as 'notches'.

There is no standard test for notch sensitivity in fatigue. However, investigations of fatigue failures have shown that some materials are much more sensitive to notches than others. In estimating the fatigue life of steel parts a common practice is to define a notch sensitivity q where:

$$q = \frac{K_F - 1}{K_t - 1}$$

and: K_t = theoretical or geometrical stress concentration factor

K_F = fatigue strength reduction factor for presence of notch

$$= \frac{\text{endurance limit of notch-free specimen}}{\text{endurance limit of notched specimen}}$$

In a particular application, K_t is determined from the geometry of the part using charts such as those supplied by the Engineering Sciences Data Unit and q is estimated from Peterson (1974). Hence K_F is found from:

$$K_F = 1 + q(K_t - 1)$$

This procedure is open to criticism because:

1. it assumes that notches of the same K_t but different geometry will have the same effect on fatigue life; and
2. it ignores the type of loading, whether it is reversed axial loading or reversed bending. In cases of doubt recourse should be had to experimental S/N curves for notched specimens. If this is not possible a conservative design approach is to put $K_F = K_t$.

Size effect

Large specimens or components often exhibit lower values of fatigue strength than geometrically similar smaller specimens or components. This 'size effect' is particularly evident when a stress gradient occurs in the section, for example under the action of a bending moment, and in cases of non-uniform stress distribution arising from the presence of a stress concentration. The un-notched fatigue limit under axial load appears to be nearly independent of specimen size.

Mann (1967) and Shigley (1986) give information which should enable some preliminary allowance to be made for size effect in design calculations.

Experimental scatter

Scatter in experimental observations of fatigue life is inherent in materials. It arises from the critical dependence of crack initiation on the maximum local stress and the microscopic properties of the material.

The statistical distribution of the fatigue lives of a group of apparently identical specimens under given test conditions can be described by either the *log-normal* or *Weibull* distributions (see Horger, 1965; Osgood, 1970). It is now the usual practice to present fatigue data in the form of scatter bands around mean lines. For example, the Engineering Sciences Data Sheets on fatigue give scatter bands which enclose 90 percent of the results and exclude the top 5 percent and the bottom 5 percent. The designer then has some indication of the reliability of the data being used.

Application to design

For a particular application S'_e is known. Allowance is made for the effects of surface finish and stress concentration by using factors K_S and K_F.

$$S_e = \frac{S'_e}{K_S K_F}$$

Factors may also have to be introduced to allow for adverse size effects and other influences which act to degrade S_e. Then design for the maximum axial stress not to exceed $\frac{S_e}{F_d}$ where F_d is the factor of safety selected.

10.1.3 Design of elements subject to fluctuating axial stress

The basis for design is the A/M diagram in which a curve of constant fatigue life can be represented by an equation of the form:

$$\frac{\sigma_a}{S_e} = 1 - \left(\frac{\sigma_m}{S_u}\right)^x$$

where S_u = static tensile strength.

For convenience of calculation the curves of constant fatigue life are often replaced by a straight line in the first quadrant, the modified Goodman line. This is a conservative practice for ductile materials where $x > 1$, but may be unsafe for brittle materials. However, our attention is confined to ductile materials as they are the most commonly used in engineering practice.

Data on the performance of notched specimens under fluctuating axial stress may not be available. It is then necessary to allow for the effects of surface roughness and stress concentrators by applying the factors K_S and K_F. They are usually applied to the

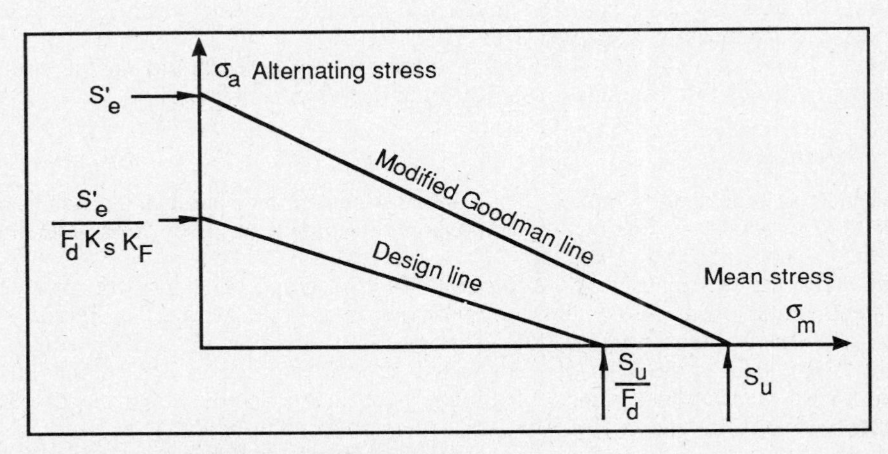

Figure 10.5 Modified Goodman diagram

alternating component of stress and not the mean stress, as the static tensile strength of ductile materials is relatively insensitive to surface finish and the effect of a stress concentration may be either to lower or to raise the tensile strength.

A design factor of safety is used to cover uncertainties in the design and the resulting diagram then becomes the basis for design (Figure 10.5).

It should be noted that the implied assumption that K_F is independent of mean stress may not be correct, and may lead to an over-optimistic design. Experimental data on the fatigue behaviour of notched specimens under non-zero mean stress should be consulted if at all possible.

10.2 Stress concentrations

The importance of stress concentrations in engineering design has already been emphasized. Any non-uniformity or inhomogeneity in a material can give rise to high local stresses and is therefore a potential source of weakness. Stress concentrations in an element may arise from:

1. change of shape;
2. application of load;
3. residual stress.

10.2.1 Change of shape

Changes of shape such as those shown in Figure 10.6 give rise to crowding of the stress trajectories (lines of force flow similar to streamlines in fluid flow). This phenomena is illustrated in Figure 10.7. Figures 10.8 and 10.9 show how high local stresses may result from the application of a concentrated load or from the presence of residual stresses remaining in the material after manufacture.

The theoretical (or geometrical) stress concentration factor is:

$$K_t = \frac{\text{maximum stress in section}}{\text{average stress in section}}$$

In the case of direct tension, $\sigma_{max} = K_t \left(\frac{P}{A}\right)$

K_t depends on the geometry of the element in the region of stress concentration. It can have high values for sharp changes of section.

The effect of a stress concentration on the strength of an element under dynamic

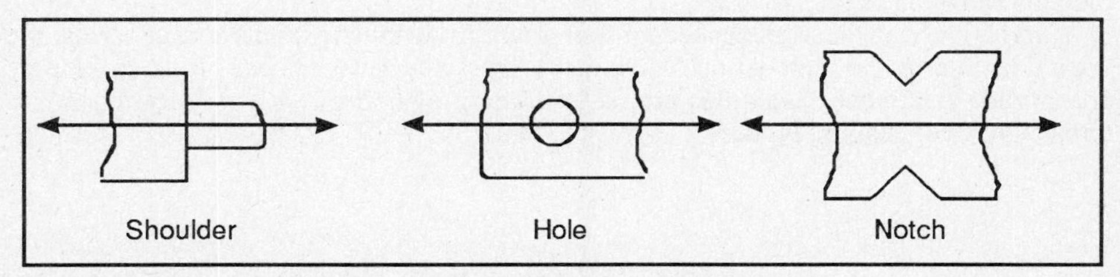

Figure 10.6 Stress concentrations due to change of shape

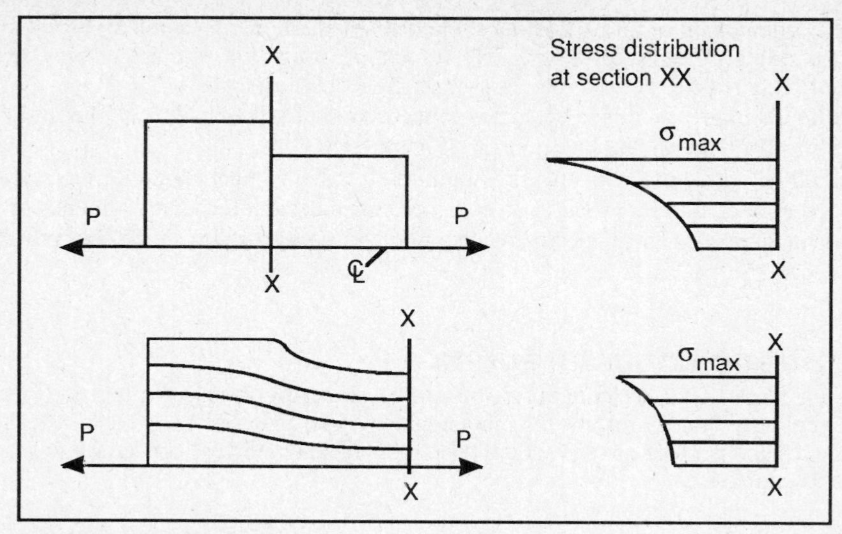

Figure 10.7 Stress trajectories

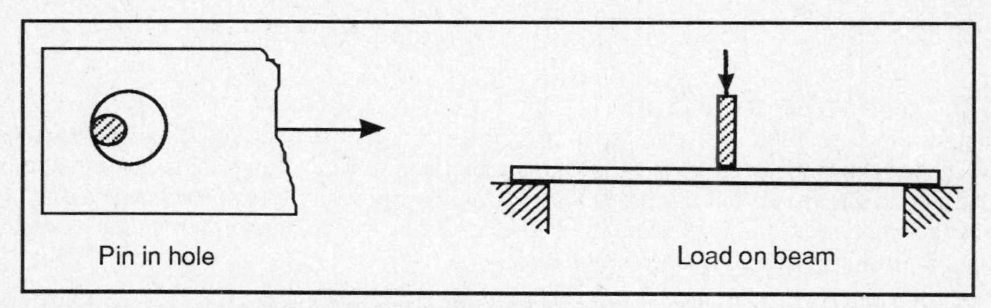

Figure 10.8 Examples of localized loading

loading has already been discussed in Section 10.1. For ductile materials under static load where plastic flow can occur in the neighbourhood of the stress concentration, the effect can be represented by a plastic stress concentration factor K_p. K_p is less than K_t; it can be related to K_t empirically; see Faupel and Fisher (1980, p. 851).

10.2.2 Residual stresses

To mitigate the adverse effects of stress concentrations there are several courses open to the designer. If the stress concentration arises from a change in geometry, he/she can arrange to make this change as gradual as possible, for example by using generous fillet radii, as shown in Figure 10.10.

Alternatively, if the stress concentration arises from some residual tensile stress, the critical region of the element can have a compressive stress induced in its surface by appropriate treatment. Examples are: shot-peening of helical springs, forming screw threads by a cold rolling process.

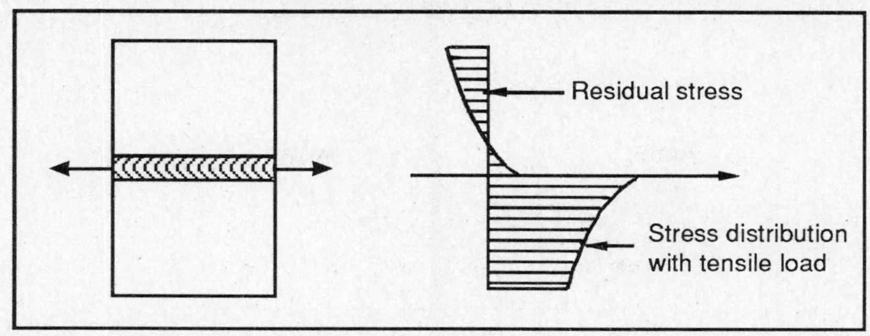

Figure 10.9 An example of a stress concentration due to residual stress in a welded plate is illustrated

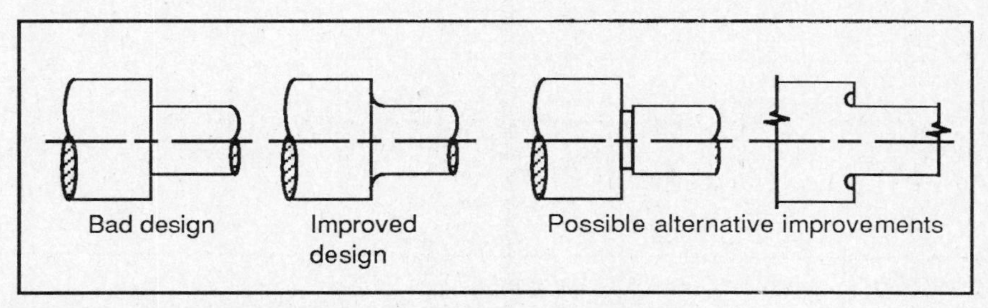

Figure 10.10 'Designing out' geometric stress concentrators

10.3 Notes from a designer's workbook

10.3.1 Design of a jack-hammer

Figure 10.11 shows a schematic sectioned sketch of a pneumatic jack-hammer . The device consists essentially of a cylinder, a pneumatic driven hammer and a tool (often a chisel) which is hammered automatically once the valve is opened. To cushion the chisel at the bottom of the stroke, where there is no resistance from the work piece, two coil springs of spring constants 0.4 MN/m (spring A) are fitted to the tie rods. Similar springs (B) are fitted to cushion the chisel at the top of the stroke. Both sets of springs are fitted without pre-loading.

Air pressure at 500 kPa is admitted into the cylinder at the start of each downward stroke of the hammer. The hammer travels 50 mm before it strikes the chisel. The air is turned off just as the chisel begins to move. The compressed air is exhausted through ports which open as the hammer passes them. The chisel and hammer both have masses of 10 kg each.

Two threaded rods are used to hold the assembly together. The nuts on the tie rods are tightened to give a pre-tension of 5 kN in each rod. The rods are hot rolled low carbon steel with S_u of 600 MPa. For hot rolled finish on the rods, K_S = 1.5 and the threads induce stress concentration for which K_F = 1.5.

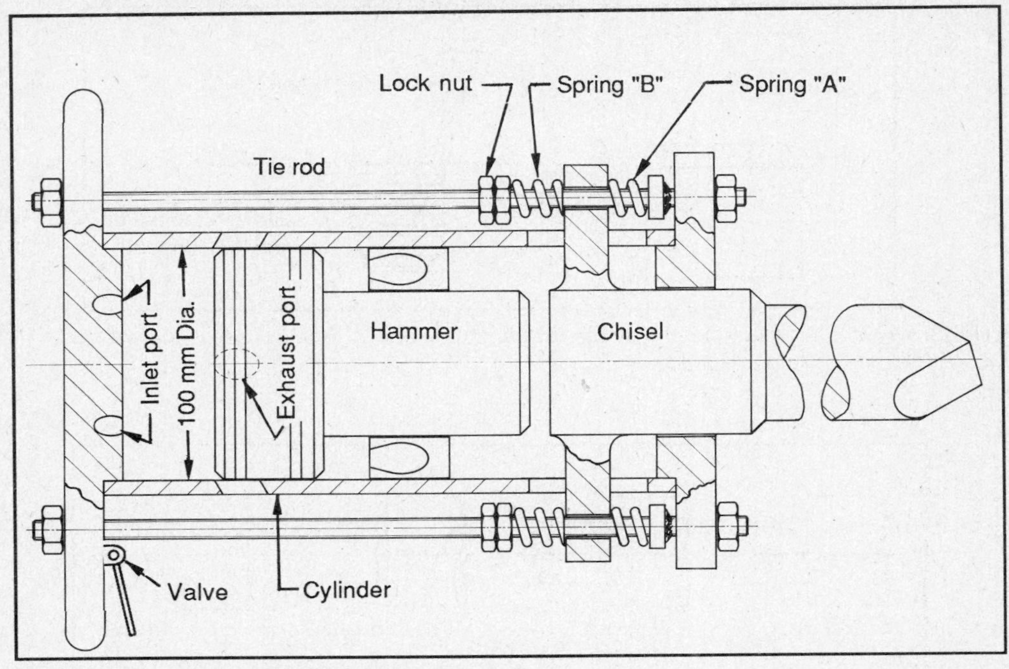

Figure 10.11 Schematic view of jack hammer

1. Assuming that the force in the tie rods fluctuates between +11.3 kN and +0.6 kN, determine the diameter of the tie rods. A design factor of safety of 2.0 may be used.
2. Determine the way in which the axial force in the tie rods varies as a function of time. Draw a graph of force against time and show that the values quoted in (1) are reasonable estimates under certain assumptions. State all assumptions.

Solution

a. Tie-rod design.

Mean Load = 6000 N
Amplitude of fluctuating load = 5300 N

$$\text{Estimate } S'_e = \frac{1}{2} S_u = 300 \text{ MPa}$$

$$S_e = \frac{S'_e}{f_d K_s K_f} = \frac{300}{2 \times 1.5 \times 1.5} = 67 \text{ MPa}$$

$$\boxed{S_e = 67 \text{ MPa}}$$

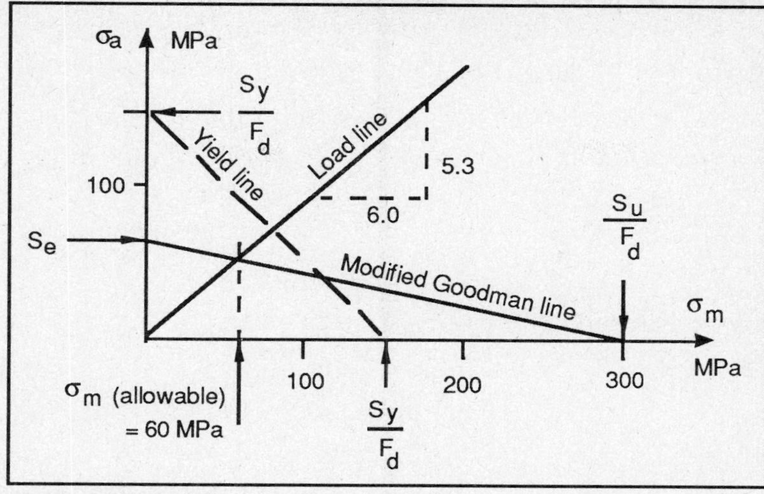

Figure 10.12 A-M diagram

Design for tension: $(\sigma_m)_{all} = (load)_{mean} \times \left(\frac{4}{\pi D^2}\right)$

$$D^2 = \frac{6000 \times 4}{60 \times 10^6 \pi}$$

This gives a value for D $= 0.0113$ m

Hence, make the diameter $D = 12mm$

b. Load – time diagram analysis

(i) loads: from assumption 1 (see below)
force on hammer after
valve is opened = pressure x area

$$= 500 \times 10^3 \times \frac{\pi}{4} \times (0.1)^2$$
$$= 3930 \text{ Newtons}$$

from assumptions 2 and 3, acceleration of hammer (a)

$$= \frac{\text{force on hammer}}{\text{mass of hammer (m)}}$$
$$= \frac{3930}{10} = 393 \text{ m/sec}^2$$

Let velocity of piston at the cut-off of air be V m/sec;
length of piston stroke to cut-off

$$= S = 0.05 \text{ m}$$

then $V^2 = 2as$
$= 2 \times 393 \times 0.05$
$V = 6.3$ m/sec

Total momentum of hammer before impact $= m_1 V$

Momentum of hammer and chisel after impact $= (m_1 + m_2)U$

where: m_2 = mass of chisel; U = velocity after impact

By conservation of momentum $= m_1 V = (m_1 + m_2)U$

$$10 \times 6.3 = 20\ U$$

$$U = 3.15\ m/sec$$

From assumption 4

Energy possessed by the hammer and chisel

after impact $= \frac{1}{2}(m_1 + m_2)U^2$

$= 0.5 \times 20 \times 3.15^2$

$= 99\ Nm$

The potential energy stored in a spring of stiffness k when deflected a distance X is given by:

$$PE = \frac{1}{2}kX^2$$

Hence at bottom of chisel stroke when there is no work piece, the potential energy stored in the spring

= kE of hammer-chisel combination

Hence, $0.5 \times 2 \times 4 \times 10^5 X_1^2$

$= 99$

$X_1 = 0.016\ m = 16\ mm$ $\boxed{X_1 = 16mm}$

Peak spring force = kX_1 (acting upward on chisel)

$= 4 \times 10^5 \times 0.016$

$= 6300\ N$

From assumption 5

upward velocity of the chisel at the return to its equilibrium position = 3.15 m/sec

At this position, springs B start to decelerate the chisel, but the hammer continues to the top of the cylinder.

At this point

K.E. of chisel $= \frac{1}{2}mU^2$

$= 0.5 \times 10 \times 3.15^2$

$= 49.6\ Nm$

P.E. of Springs (B) when chisel reaches

its highest point $= \frac{1}{2}kX_2^2$

$$= 0.5 \times 2 \times 4 \times 10^5 X_2^2 = 49.6$$

$$X_2 = 0.011 \text{ mm}$$

$X_2 = 11 mm$

Peak spring force $= kX_2 = 4400$ kN (acting downward on chisel)

from assumption 6
the chisel stabilises at or near the equilibrium position before the cycle repeats.

(ii) times:

time taken for hammer to reach the chisel (t_1)

$$= \frac{\text{velocity } V}{\text{acceleration } a}$$

$$= \frac{6.3}{393} = 0.016 \text{ sec}$$

$t_1 = 0.016$ sec

time taken for chisel to reach the end of its stroke from impact (t_2)

$$t_2 = \frac{1}{4}\left(2\pi\sqrt{\frac{m_1}{k}}\right)$$

$$= \frac{1}{4}\left(2\pi\sqrt{\frac{10}{4 \times 10^5}}\right) = 0.008 \text{ sec}$$

$t2 = .008$ sec

{Note: natural frequency of vibration $= \frac{1}{2\pi}\sqrt{\frac{k}{m}}$ }

similarly, time taken for chisel to return to equilibrium position
(t_3) = 0.008 sec

$t_3 = .008$ sec

time taken for hammer to return to the top of the cylinder (from assumption 7) $= t_4$

$$= \frac{2 \times .05}{3.15} = 0.032 \text{ sec}$$

$t_4 = .032$ sec

Approximate cycle time

$$= t_1 + t_2 + t_3 + t_4$$

$$= 0.07 \text{ sec (approximately 15 Hz)}$$

Assumptions and load-time diagram:

1. Cylinder pressure = 500 kPa. Although the actual pressure will be less than 500 kPa, due to losses through the valves and inlet ports, the design which results from this assumption will be conservative.
2. Cylinder pressure is constant over the full stroke of hammer.This assumes that the air to the cylinder is supplied at a sufficient rate to maintain the pressure. As the piston accelerates, the valve will be opening further, so this assumption may be valid. The resulting design will be conservative, because the pressure would probably be dropping.
3. Friction in the cylinder is negligible. Normally the contact force of the sealing rings provide for a nominally small friction force (perhaps 10 N). Since this is several orders of magnitude less than the external force, the assumption is justified.
4. The gas pressure in the cylinder after impact between the hammer and chisel is negligible. Since the exhaust ports are large and open to the atmosphere, this assumption approximates to the true loading condition. However, the actual force on the hammer is greater than that estimated by the analysis following this assumption, and a safety factor should be incorporated in the design.
5. When there is no workpiece under the chisel, there is no loss of energy from the system after the hammer has the chisel. Hence the chisel could vibrate about its equilibrium position between springs A & B. This assumes that frictional losses are negligible.
6. There is sufficient damping in spring B to bring the chisel to

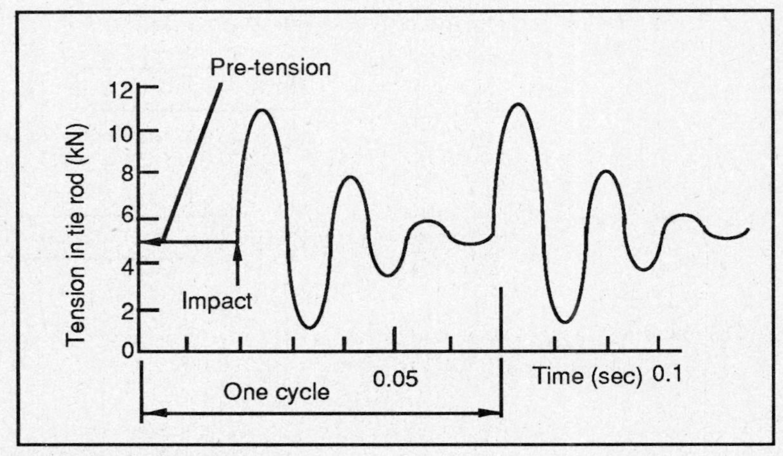

Figure 10.13 Load – time diagram

near-rest in the equilibrium position before the hammer strikes it on each cycle.

7 The hammer decelerates uniformly from the time it leaves contact with the chisel until it reaches the top of the cylinder. This assumes that the trapped air in the cylinder is at constant pressure (not true in practice).

10.4 Exercises in design for dynamic loads

10.4.1 Column for riveting machine

Figure 10.14 shows a schematic sketch of a small riveting machine. You are currently employed by a machine tool manufacturer tendering for the production of these machines. The head stock of the machine incorporates a pneumatic clamp for holding the work-piece in position during the riveting cycle. A typical cycle of operation is as follows:

1. work is placed on machine support for clamping;
2. the clamp is brought down into contact with the work-piece and clamping pressure is applied. Clamping load may vary up to 8.0 kN;
3. a burst of impacts are applied during the riveting process at approximately 10 hz. The riveting impact load is a function of air pressure, with a maximum estimated at 16.0 kN;
4. on completion the clamp and tool are raised ready for the next cycle.

You are to design the main column of the machine, assuming the plant will produce approximately 1000 machines per batch. A design factor of safety of 2.0 may be used.

The available materials are:

hot rolled carbon steel with $S_u = 600$ MPa;
cold drawn carbon steel with $S_u = 1500$ MPa;

Table 10.1 Fatigue strength modifying factors for steels

Type of finish	K_S	
	S_u = *600m MPa*	S_u = *1500 MPa*
Polished	1.0	1.0
Ground	1.11	1.11
Machined	1.25	1.59
Hot rolled	1.54	2.86
Forged	2.0	4.0

(i) List modes of failure for the machine column.
(ii) Determine column diameter to resist fatigue failure.
(iii) Select the best material to suit the design.
(iv) Produce a dimensioned sketch of the column.

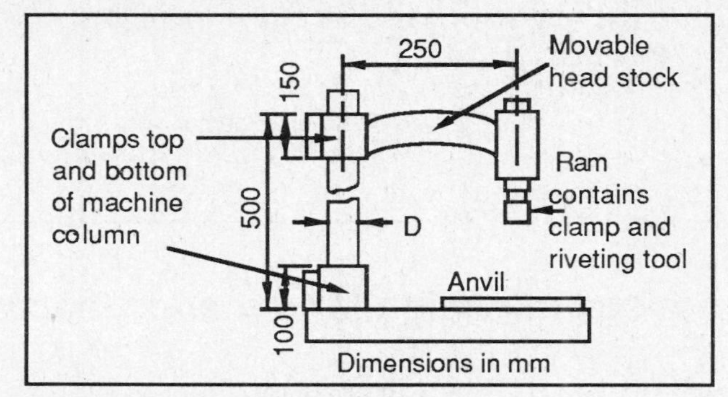

Figure 10.14 Schematic sketch of riveting machine

10.4.2 Link

A link is required for transmitting a time varying force between 22.3 kN compression to 116.6 kN tension. At each end of the link pin joints will be used to connect the link to its neighbouring machine member. The nominal distance between the pin centers is 300.00 mm, and this must be kept within the limits of ± 0.15 mm. Space restrictions limit one dimension of the link (its thickness parallel to the center lines of the pins) to a dimension of 30 mm.

The material to be used is low carbon steel with S_u = 592 MPa; S_y = 360 MPa; S_e = 243 MPa; and Young's modulus of 208 GPa.

The diameter of the pins is to be 38 mm, and the fit between link and pin is to be a clearance fit with a maximum diametral clearance of 0.05 mm and a minimum of 0.013 mm.

A design factor of safety of 3.0 may be used.

1. List modes of failure for the link.
2. Determine all dimensions for the link.
3. Make a dimensioned sketch of the link for production.

10.4.3 Bell-crank

Design of a bell-crank. Figure 10.15 shows a bell-crank lever subject to reversed loading. You are to design the crank. Pin joint sizes are noted in Table 10.2. The aims of this assignment are:

1. to clarify the concept of 'mode of failure';
2. to give practice in designing engineering components to resist failure when the applied load varies with time;
3. to develop the ability to communicate engineering information by means of dimensioned sketches.

1. List the modes of failure of the lever.
2. Specify all the key dimensions of the lever, justifying your specifications with suitable calculations (batches of 100 levers to be made).
3. What effect will it have on the section area of the lever if a force of 1.2 kN is added to F_1?

Available materials are:

(i) hot rolled low carbon steel (cost = $V /tonne);
tensile strength 600 MPa (S_y = 400 MPa);

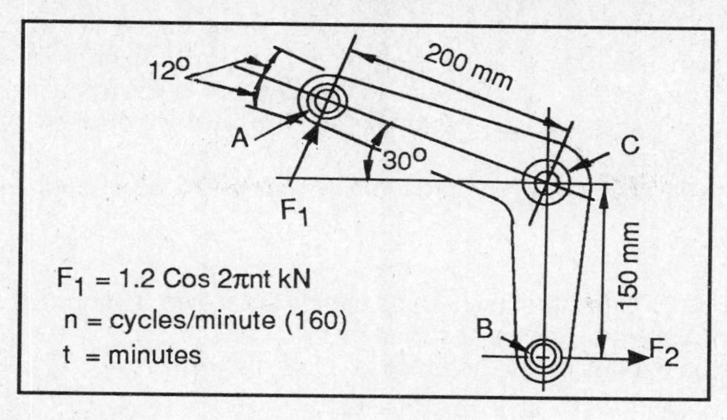

Figure 10.15 Bell-crank

(ii) cold drawn low alloy steel (cost = $3V/tonne); tensile strength 1500 MPa (S_y = 1200 MPa).

K_s values may be used from Table 10.1. Design Factor of Safety $F_d = 2$ is to be used.

Further data

Table 10.2 Sizes of nylon bushes in bell crank

Location	*Nylon bush*	
	Outside diameter (mm)	*Length (mm)*
A	23	30
B	23	30
C	28	30

The crank oscillates through $\pm$ 12° about the position shown.
The lever support is at 'C' and forces F_1 and F_2 are transmitted at A and B.

10.4.4 Torsion bar

The front suspension of a popular sports car of the sixties and seventies is shown simplified in the sketch. Essentially the torsion bar acts as a spring. The torsion bar is

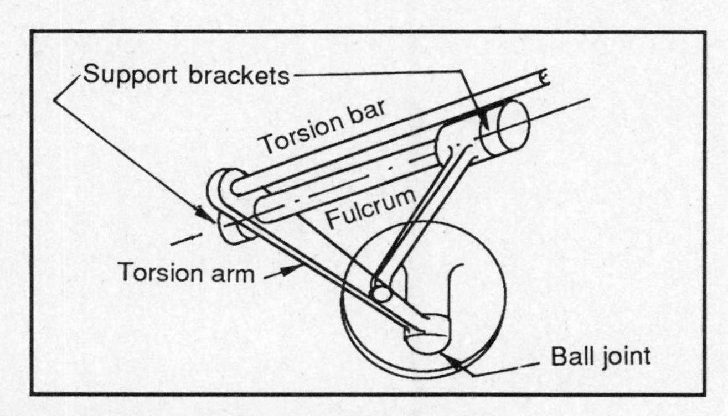

Figure 10.16 Schematic front suspension of automobile

fixed to the arm by splines and fixed to the car body at a point 1.9 m away. The distance from the lower ball joint assembly to the center line of the fulcrum is 450 mm. The distance from the fulcrum to the center of the torsion bar is 150 mm. The brackets are fixed to the chassis. The car weighs 1.2 tonnes and has a weight distribution between the front and back of 60/40.

Your task is to design the arm and torsion bar to resist fatigue failure.

Notes

1. The fulcrum is 100 mm in diameter and should bear over a length of 30 mm in the arm. The arm is initially chosen as rectangular.
2. Both the torsion bar and arm to be made of a low alloy steel with S_u = 1400 Mpa and S_y = 1000 Mpa; G = 80 GPa; E = 210 GPa.
3. The torsion bar is available in the cold drawn state and the arm as hot rolled. The appropriate K_s values are:

1.0	polished
1.11	ground
1.59	cold drawn/machined
2.7	hot rolled

4. The usual extra load due to driving over a large bump at moderate speed is estimated as being equal to 3/4 of the static load.

The problem

1. Comment on the appropriateness of using a modified Goodman diagram approach for this exercise.
2. Determine the dimensions of the arm to resist fatigue for an infinite life; determine the surface finish.
3. Improve the design of the arm.
4. Determine the diameter of the torsion bar for an infinite fatigue life. Design for bending only. As far as you are able, justify this.

The fitting of the torsion bar to the arm may be taken as having no slip. Use Fd = 2 (is this reasonable?) and K_t = 1.4. State all assumptions, and attempt to justify these.

Chapter 11

Elements subject to axial, transverse and torsional loads

An enemy had once told him ... he needed more buckle and less swash.
Julian Mitchell

In engineering design we need more swash and less buck
AES

Concepts introduced	buckling of columns; end conditions and slenderness ratios; combined torsional and transverse load fatigue of shafts.
Methods presented	Euler buckling theory; empirical curve fit to short column behaviour; tangent modulus procedure for plastic buckling of short columns; formulae for shaft design.
Applications	pin jointed triangular truss design; machine tool cutter shaft design.

11.1 Elements in tension

We consider the case of a bar (cross-section area A) subject to a steady axial load (P), and assume that the material is homogeneous and that the stress distribution across any section is uniform. Conventional notation is used throughout with the suffix 'all' standing for 'allowable', that is, acceptable upper limit for some critical stress or strain.

Modes of failure: (a) yielding (*ductile materials*) or fracture (*brittle materials*); (b) excessive deflection.

Mathematical model:

(a) $$\sigma = \frac{P}{A}$$

(b) $$\delta = \frac{Pl}{AE}$$

Factor of safety:

$$\sigma \leq \sigma_{all} = \left(\frac{S_y}{F_d}\right) \text{(yielding)}$$

$$\sigma \leq \sigma_{all} = \left(\frac{S_u}{F_d}\right) \text{(fracture)}$$

Design inequalities: (a) $A \geq \frac{P\,F_d}{S_y}$

$$A \geq \frac{P\,F_d}{S_u}$$

(b) $A \geq \frac{P\,1}{E.\delta_{all}}$

11.2 Elements in compression

11.2.1 Long (or slender) columns

We consider the design of columns — relatively long compression members. In a typical case the design problem is to determine the cross-sectional dimensions of a column given:

1. the magnitude of the compressive load P;
2. the physical length L; and
3. the method of supporting the ends of the column.

Mode of failure

Columns fail by buckling — a form of elastic instability. It is a physical impossibility to load a column so that the line of action of the compressive force passes through the center of its cross-section. Moreover, due to manufacturing inaccuracy a column will not be perfectly straight. As a result of these effects any section of a column is subject to a bending moment, and if the compressive force is sufficiently high ($P = P_{cr}$ say) it will buckle sideways.

Mathematical model

I = second moment of area of the column cross-section about an axis perpendicular to plane of buckling.

$I = Ar^2$, where r = radius of gyration. **(11.1)**

l = 'effective length' of column, that is, the half wavelength of the deflection curve of the column (assumed to be a sine wave).

For column with pinned ends . $l = L$
For column with built-in ends $L = L/2$

The ratio of effective length 'l' to actual length 'L' of a column is a critical parameter in predicting column performance and it is fully determined by the 'end conditions' of the column. Figure 11.1 shows schematically the deflection behaviour of columns with various end conditions.

For long slender columns Leonhardt Euler derived an equation for predicting the critical buckling load P_{cr}:

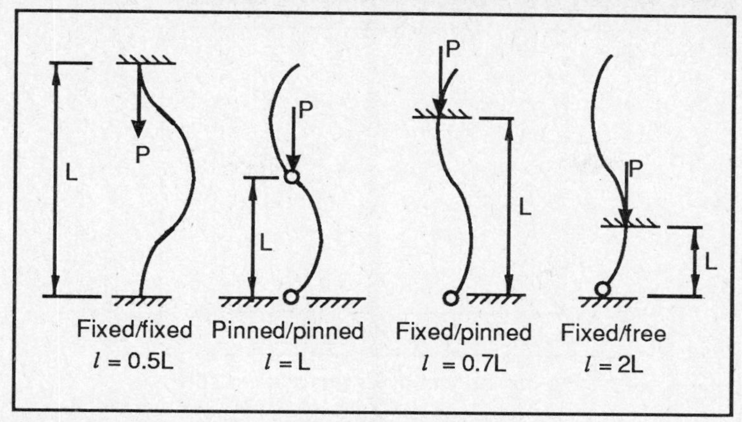

Figure 11.1 Buckling behaviour with various end conditions

$$P_{cr} = \frac{\pi^2 E I}{l^2} \tag{11.2}$$

Substituting from equation 11.1:

$$\frac{Pcr}{A} = \frac{\pi^2 E}{\left(\frac{l}{r}\right)^2} \tag{11.3}$$

$\frac{P_{cr}}{A}$ is often called 'column stress', since the ratio has the dimensions of stress and

$\frac{l}{r}$ is called the 'slenderness ratio'.

11.2.2 Short columns

A number of empirical and semi-empirical equations have been developed to predict the behaviour of short columns. R. J. Roark and W. C. Young in their book *Formulas for Stress and Strain* (McGraw-Hill) review a wide range of possible design relations, including the Johnson parabola described below. This book provides a comprehensive tabulation of results of column load per unit area of cross-section for a variety of materials ranging from steel to timber.

Short columns may suffer inelastic buckling at failure. For mild steel the minimum slenderness ratio corresponding to a Euler column stress is approximately 100. Below this slenderness ratio some inelastic deformation will occur at high loads. One technique for predicting column behaviour under these conditions is the *tangent modulus* approach, see *'Aircraft Structures for Engineering Students'* by T. H. G. Megson (Edward Arnold). In this approach the following equation is used when the material of construction is outside its elastic range:

$$\sigma_c = \frac{P_{cr}}{A} = \frac{\pi^2 E_t}{\left(\frac{l}{r}\right)^2} \tag{11.4}$$

where E_t is the slope of the local tangent to the stress-strain curve for the material, see Figure 11.2.

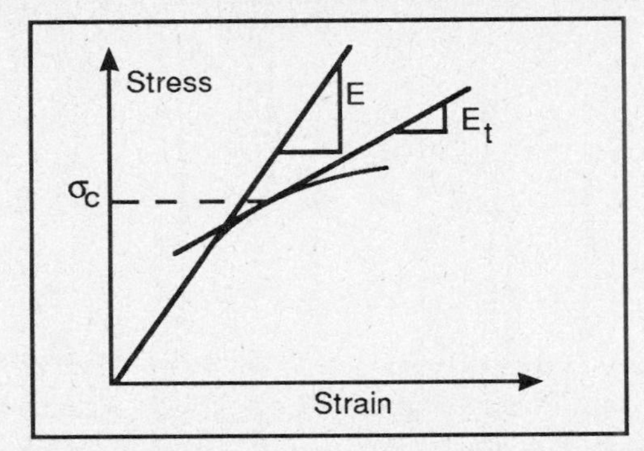

Figure 11.2 Determining the tangent modulus for short columns

As Megson has pointed out, experiments with aircraft structures have demonstrated the validity of this method for predicting the performance of compression members.

In structural engineering, a number of equations have been proposed, a popular one being the Johnson parabola.

$$\frac{P_{cr}}{A} = S_y\left\{1 - \left(\frac{S_y}{4\pi^2 E}\right)\left(\frac{l}{r}\right)^2\right\} \tag{11.5}$$

When $\frac{P_{cr}}{A}$ is plotted against $\frac{l}{r}$, the Johnson parabola is:

1. symmetrical about the vertical axis and intercepts it at $\frac{P_{cr}}{A} = S_y$ and
2. tangential to the Euler curve.

From analytical geometry, the point of tangency is found to be the point where:

$$\frac{P}{A} = \frac{S_y}{2} \; ; \; \frac{l}{r} = \sqrt{\frac{2\pi^2 E}{S_y}}$$

The results of the mathematical modelling of column behaviour are displayed in Figure 11.3 where column stress at failure is plotted against slenderness ratio. Figure 11.3 also indicates the application of a factor of safety to obtain design values of P/A as explained in the next section.

11.2.3 Application to design

Factor of safety

In column design a factor of safety is applied to the axial compressive load:

$$P \le P_{all} = \frac{P_{cr}}{F_d}$$

F_d varies from about 1.7 for short columns to values greater than 2 for long columns. Columns with $\frac{l}{r} > 180$ are rare.

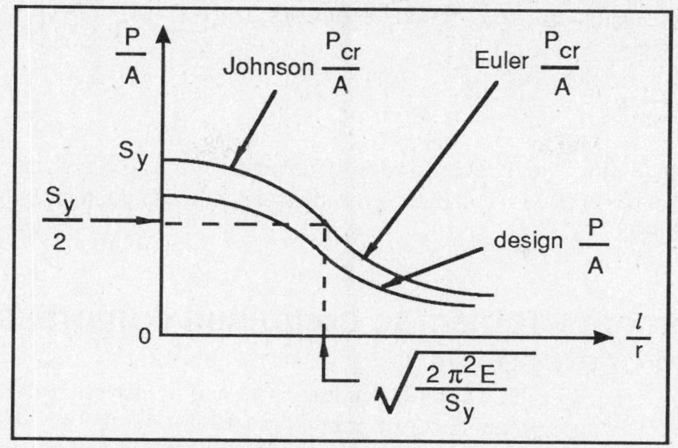

Figure 11.3 Column stress as function of slenderness ratio

Design inequalities

Euler: $$\frac{l}{r} > \sqrt{\frac{2\pi^2 E}{S_y}}$$

$$P \le P_{all} = \frac{1}{F_d} A \frac{\pi^2 E}{\left(\frac{l}{r}\right)^2} \tag{11.6}$$

Johnson: $$\frac{l}{r} < \sqrt{\frac{2\pi^2 E}{S_y}}$$

$$P \le P_{all} = \frac{1}{F_d} A \ S_y \left\{1 - \frac{S_y}{4\pi^2 E}\left(\frac{l}{r}\right)^2\right\} \tag{11.7}$$

The design inequalities are often expressed in terms of the column stress:

Euler: $$\boxed{\sigma_c \le \sigma_{c\text{-}all} = \frac{1}{F_d} \frac{\pi^2 E}{\left(\frac{l}{r}\right)^2}} \tag{11.8}$$

Johnson: $$\boxed{\sigma_c \le \sigma_{c\text{-}all} = \frac{1}{F_d} S_y \left\{1 - \frac{S_y}{4\pi^2 E}\left(\frac{l}{r}\right)^2\right\}} \tag{11.9}$$

Selection of materials and dimensions

The design procedure is then as follows for given P and l:

1. Choose material. ⇨ $\boxed{\sqrt{\frac{2\pi^2 E}{S_y}}}$

2. Make trial decision on cross-section dimensions. ⇨ $\boxed{r}$ ⇨ $\boxed{\frac{l}{r}}$

⇨ $\boxed{\left(\frac{P}{A}\right)_{all}}$ ⇨ $\boxed{P_{all}}$

3. Compare P_{all} with P and revise design if necessary. If $P_{all} > P$, the design is safe, perhaps oversafe. If $P_{all} < P$, the design is unacceptable and another, larger value of r has to be tried.

11.3 Elements subject to combined compression and bending

Let: σ_{b-all} = allowable stress in bending, obtained by dividing yield strength by appropriate factor of safety.

σ_{c-all} = allowable column stress, obtained from Euler or Johnson equation and an appropriate factor of safety.

We ignore shear stresses due to bending and apply the following design inequality:

$$\boxed{\left(\frac{\sigma_b}{\sigma_{b\text{ - }all}} + \frac{\sigma_c}{\sigma_{c\text{ - }all}}\right) \leq 1}$$

11.4 Elements subject to transverse loads — beams

The most likely mode of failure of an element subject to transverse load is *excessive deflection*, although stresses must always be checked.

As long as the method by which a beam is supported is known, together with the type of restraint at these supports, then theory from the mechanics of solids can be used to predict deflections quite accurately. Failure due to excessive deflection, while embarrassing, is not catastrophic — the result may be unsightly or unnerving to the spectator, there may be cracking of partitions in a building, but there is little or no danger to life and limb. For these reasons it is common to use a factor of safety of close to unity in calculations of deflections.

At any cross-section of a beam, theory predicts maximum stress and strain at the outermost fibres of material which are furthest from the neutral axis. Material placed far from the neutral axis is thus more effective in resisting bending; hence the use of I-section beams in buildings and bridges.

The shafts of many rotating machines such as motors, pumps and fans are subject to both bending and torsion. Usually they have to be designed so that their maximum deflection is within acceptable limits and conforms to requirements for air gaps and fine running clearances between rotors and stators.

Beams in steel structures in buildings and factories are designed in accordance with the requirements of Standard codes of design practice. An example is AS 1250 *The Use of Steel in Structures* (see Section 5 and Appendix A of that publication).

11.5 Elements subject to bending and torsion — shafts

We consider the design of shafts to transmit torque. Shafts are also subject to bending moments resulting from gear, pulley, out-of-balance and other transverse forces, and we have therefore to take into account the combined effect of torsion and bending. Axial forces may also be present if helical or bevel gears are used.

The stresses at a point on the surface of a shaft fluctuate cyclically as the shaft rotates, so that one important mode of failure is fatigue fracture under combined stress. A fatigue fracture, if it occurs, will initiate from a stress concentration on the surface of the shaft such as a keyway, sharp fillet, or even a rough machining mark. The stress system is biaxial; the principal stresses σ_1 and σ_2 and maximum shear stress τ_{max} all vary sinusoidally with time, thus :

$$\boxed{\sigma = \sigma_m + \sigma_a \cos(\omega t)}$$

11.5.1 Modes of failure

Shafts may fail by other ways besides those already mentioned. For example, if the slope of a shaft at its support bearing is too high, the operation of the bearings may be adversely affected and their lives shortened. A long shaft may suffer from excessive angular twist. A shaft is an elastic body and has a natural frequency of vibration. If this frequency is close to its operating speed then damagingly large vibrations will ensue. To sum up, we have to design shafts to resist some or all of the following modes of failure:

1. fracture by fatigue;
2. excessive lateral deflection;
3. excessive slope at bearings;
4. excessive angular twist;
5. excessive vibrations (critical speeds);
6. buckling under axial load.

The following discussion is concerned with the first mode of failure in order to illustrate some procedures for designing against fatigue.

11.5.2 Prediction of failure by fatigue

In a typical design case we know the stresses at a point on the surface of a shaft of diameter d at a section where the bending moment is M and torque is T ($r = \frac{d}{2}$). The principal strèsses and maximum shear stress can then be found as shown in Figure 11.4.

The maximum bending stress is σ_b where:

$$\sigma_b = \frac{M\,r}{I} = \frac{32M}{\pi d^3}$$

and the shear stress is τ where:

$$\tau = \frac{T\,r}{I_p} = \frac{16T}{\pi d^3}$$

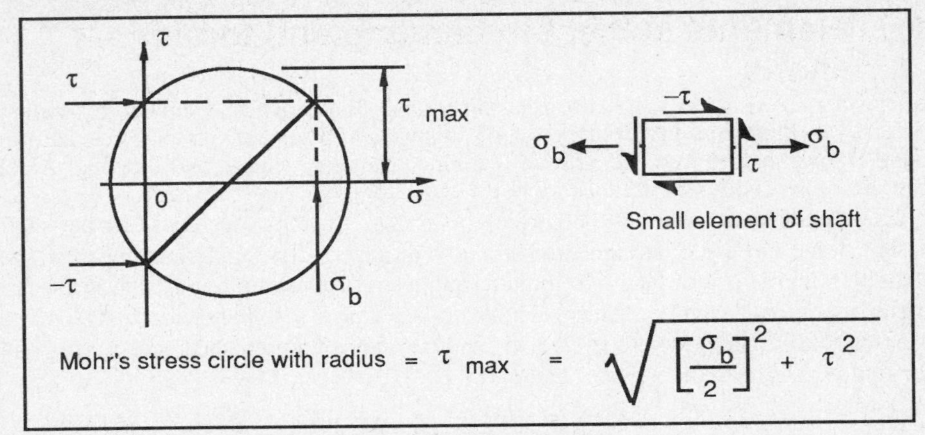

Figure 11.4 Finding the maximum shear stress in a shaft

The principal stresses and maximum shear stress are:

$$\sigma_1 = \frac{\sigma_b}{2} + \sqrt{\left(\frac{\sigma_b}{2}\right)^2 + \tau^2} = \frac{16}{\pi d^3}\left(M + \sqrt{M^2 + T^2}\right)$$

$$\sigma_2 = \frac{\sigma_b}{2} - \sqrt{\left(\frac{\sigma_b}{2}\right)^2 + \tau^2} = \frac{16}{\pi d^3}\left(M - \sqrt{M^2 + T^2}\right)$$

$$\sigma_3 = 0$$

$$\tau_{max} = \sqrt{\left(\frac{\sigma_b}{2}\right)^2 + \tau^2} = \frac{16}{\pi d^3}\sqrt{M^2 + T^2}$$

Both the maximum shear stress theory and the maximum shear strain energy theory are used in practice to predict failure by fatigue fracture. The simple approach is to base the calculations on peak values of stresses rather than try to adapt the A-M diagram (previously established for fluctuating uniaxial stress) to deal with fluctuating combined stresses.

Thus the maximum shear strain energy theory predicts failure when:

$$(\sigma_1 - \sigma_2)^2 + (\sigma_2 - \sigma_3)^2 + (\sigma_3 - \sigma_1)^2 = 2 S_e^2$$

which reduces to: $$(\sigma_1)^2 + 3\tau^2 = S_e^2$$

that is: $$\frac{32}{\pi d^3}\sqrt{T^2 + \frac{3T^2}{4}} = S_e$$

This is a theoretical prediction, which does not allow for the effects of stress concentrations and the other factors noted in Section 10.1.2 as affecting the fatigue strength of a component.

11.5.3 Application to design

Australian Standard

The notation used in the Australian Standard for the design of rotating steel shafts, AS 1403, is as follows:

D = shaft diameter
F_S = factor of safety
K_S = size factor
K = stress-raising factor[1]
M_q, F_q, T_q = maximum bending moment, axial tensile force and torque respectively at the cross-section of the shaft under consideration

We continue to denote the endurance limit of the shaft material as S_e, based on laboratory rotating bending tests of polished speciments 8-10 mm in diameter.

The design inequalities recommended by Borchardt (1973) and AS 1403 are as follows:

1. Fully reversed torque (general case):

$$D^3 \geq \frac{10F_sK_sK}{S_e}\sqrt{\left(M_q + \frac{F_qD}{8}\right)^2 + \frac{3T_q^{\,2}}{4}} \tag{11.10}$$

2. Pulsating torque, where the numerical value of torque varies between zero and a maximum:

$$D^3 \geq \frac{10F_s}{S_e}\sqrt{\left[K_sK\left(M_q + \frac{F_qD}{8}\right)\right]^2 + \frac{3T_q^{\,2}(1 + K_sK)}{16}} \tag{11.11}$$

3. Steady torque:

$$D^3 \geq \frac{10F_s}{S_e}\sqrt{\left[K_sK\left(M_q + \frac{F_qD}{8}\right)\right]^2 + (0.87\,T_q)^2} \tag{11.12}$$

Notes on design procedure

1. Checks on fatigue strength must be made at all cross-sections of the shaft considered by the designer to be possible candidates for failure.
2. Consistent units must be used throughout, for example, kilograms, meters, seconds, newtons, pascals.
3. A low factor of safety may be used (F_s = 1.2) if peak loads are known accurately. Most prime movers are capable of producing torques far in excess of nominal values, for example, pull-out torques for electric motors may reach three to five times their nominal values. High torques during acceleration or deceleration help to to overcome inertia, but they may overload shafts. Any doubts about peak values must be reflected in high safety factors.

---

1 This factor K is equivalent to K_F the fatigue strength reduction factor due to the presence of notches, introduced in Chapter 10.

4. Highly stressed regions in a shaft should have a smooth surface finish with even changes in geometry and avoidance of sharp corners.
5. AS 1403 contains charts for determination of K_s and K, covering keyways, grooves, shoulders, shrink fits and press fits.
6. The figure of 10 on the right-hand side of the inequalities is a simple approximation of the ratio $\frac{32}{\pi}$.

Other design inequalities

The recommendation of Baumeister (1978) is based on similar reasoning. However, a fatigue strength reduction factor K_{tf}[2] is explicitly associated with the effect of any stress concentrations on fatigue life in rotating bending.

$$D^3 > 2.17\,F \sqrt{0.422\left(\frac{T}{S_y}\right)^2 + \left(\frac{K_{tf}M}{S_e}\right)^2} \tag{11.13}$$

This inequality applies to cases where T and M do not vary with time. F is the factor of safety, and because the shear stress τ is constant (does not vary as the shaft rotates), the denominator of the first term under the square root is S_y and not S_e.

$$K_{tf} = q\,(K_t - 1) + 1$$

where: q is notch sensitivity; and
K_t is the theoretical stress concentration factor in rotating bending.

Since σ_1, σ_2 and τ_{max} fluctuate cyclically with time as the shaft rotates, a more rational approach would be to extend the A-M diagram to deal with combined stresses. The application of both theories of failure to the A-M diagram is discussed by Shigley (1986), Sections 7-15 and 7-16. The design equations in the ninth edition of Marks Standard Handbook for Mechanical Engineers are based on this work (Avallone and Baumeister, 1987).

11.6 Notes from a designer's workbook

11.6.1 Design of a simple truss

A simple triangular structure is required to support a load W. Figure 11.5 shows such a truss schematically.

1. Design the structure [members (1) and (2)] for minimum weight.
2. Design it for minimum cost.

Data

l = 5 m W = 140 kN

α = variable (< 75°) F_d = 2.5

2 K_{tf} is also the same as K_F referred to in Chapter 10.

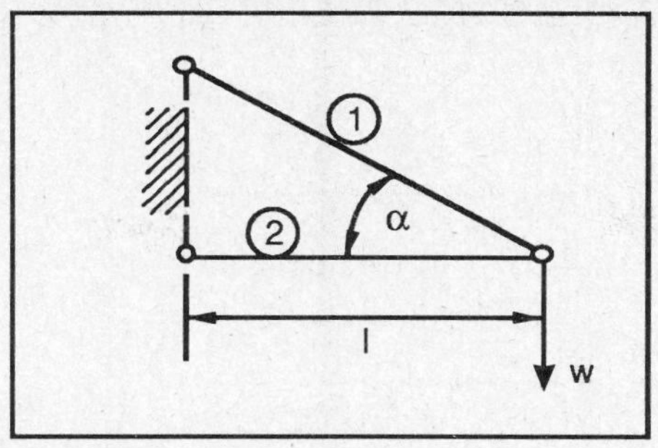

Figure 11.5 Pinned truss

Available materials (tubular sections as shown in Table 11.1)

Steel: $S_y = 300$ MPa

Aluminum: $S_y = 241$ MPa

Table 11.1 Section properties and slenderness ratio

Outside diameter O.D.mm	*Thickness* t mm	*Radius of gyration* $\sqrt{\frac{I}{A}}$ mm	*Section area* A(mm^2)	*Slenderness ratio* $\frac{l}{r}$ mm
90	6	29.8	1583	168
100	5	33.6	1492	149
110	6	36.8	1960	136
110	8	37.6	2652	133
150	8	50.3	3569	99.4
150	5	51.3	2278	97.5
160	5	54.8	2435	91.2
160	6	54.5	2903	91.7
175	5	60.1	2670	83.2

Solution

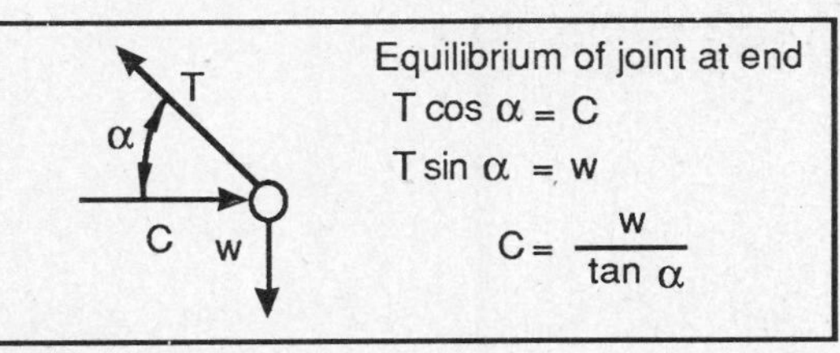

Figure 11.6 Eqilibrium of end joint

(a) Member (1); assume slender column (check later) – must not buckle.

Euler (equation 11.3); pin joint ends: $\dfrac{W}{\tan\alpha} \le \dfrac{\pi^2 E A_1}{\left(\frac{l}{r}\right)^2}$

$$A_1 \geq \frac{W\left(\frac{l}{r}\right)^2}{\pi^2 \, E \tan \alpha}$$ (a) A_1

(b) Member (2): tension – must not yield

$$\frac{T}{A} \leq S_y \, ; \quad \frac{W}{\sin \alpha} \leq A_2 S_y$$

$$A_2 \geq \frac{W}{S_y \sin \alpha}$$ (b) A_2

$$\text{Mass} = M_1 + M_2 = \rho \, l \, A_1 + \frac{\rho \, l \, A_2}{\cos \alpha}$$

$$M_{min} = \rho W l \left[\frac{\left(\frac{l}{r}\right)^2}{\pi^2 E \tan \alpha} + \frac{1}{S_y \cos \alpha \sin \alpha}\right]$$ (c) M_{min}

(c) Choice of best α (this is independent of material choice).

M is proportional to $\left(\frac{K_1}{\tan \alpha} + \frac{K_2}{\sin 2\alpha}\right)$

$$\frac{K_1}{K_2} = \frac{\left(\frac{l}{r}\right)^2 S_y}{\pi^2 E}$$; for practical values of the order of unity

Differentiating M and setting $K_1 = K_2 = 1$;

$$\frac{dM}{d\alpha} = \frac{\sec^2 \alpha}{\tan^2 \alpha} + \frac{2}{\tan 2\alpha} = 0$$

gives $\alpha = 60^o$ $\alpha = 60^o$

(d) Material choice :

buckling M_{min} is determined by the lowest value of $\frac{\rho}{E}$;

tension M_{min} is determined by the lowest $\frac{\rho}{S_y}$;

ρ = density of material (= specific gravity (SG) x 1000)

R_c = relative cost

Clearly the best choice for the compression member is steel, since both aluminum and steel have similar values for $\frac{\rho}{E}$ and the cost of aluminum is about eight times that of steel.

	S.G.	R_c	E(GPa)	S_y(MPa)	$\frac{\rho}{E} x10^8$	$\frac{\rho}{S_y} x10^5$
Steel	7.84	1	208	300	3.77	2.61
Aluminum	2.7	8	73	241	3.7	1.12
Magnesium	1.8	12	45	260	4.0	0.69
Titanium	4.6	26	117	750	3.8	0.61

Point of tangency between Euler and Johnson parabola :

$$\frac{l}{r} = \sqrt{\frac{2\pi^2 E}{S_y}}$$

Steel : $\frac{l}{r} = 117$

Aluminum: $\frac{l}{r} = 77.3$

These values of $\frac{l}{r}$ show that all available tubes (refer Table 11.1) may be regarded as slender columns for buckling.

Comparing costs – The only gain to be obtained is for the tension member where the value of $\frac{\rho}{S_y}$ for the more expensive materials is significantly lower than that for steel.

For general comparison use $\frac{\rho}{S_y} \times R_c$

Steel	2.61
Aluminum	8.96
Magnesium	8.28
Titanium	15.86

Hence, for tensile members there is some gain in choosing Mg over Al, but only in cost v. weight choice.

Best material to resist buckling — Aluminum

Best material for cost — Steel

Lightest structure:	Member (1)	Al
	Member (2)	Ti
Cheapest structure:		Steel

11.6.2 Machine shaft design

Figure 11.7 shows the cutter shaft of a milling machine. The input power is 10 kW at 4500 rpm. During the cutting cycle the material may be presented to the cutter from any direction tangential to the cutter. The design should cater for heavy shock loading and suddenly applied load.

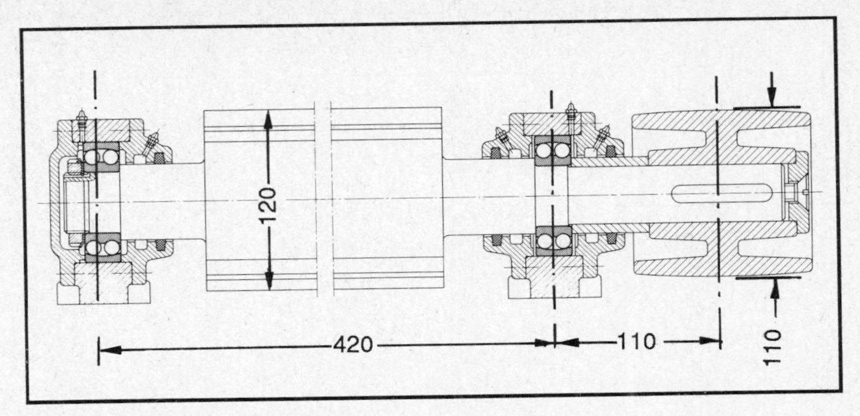

Figure 11.7 Machine-tool cutter shaft

Design a shaft capable of transmitting the applied power together with the restriction that the maximum deflection at the center of the cutter is to be less than 0.25 mm during the cutting cycle.

Clearly state how you estimated values for necessary design data not already given.

Assumptions and prior decisions

1. Assume cutting load and pulley load are point loads applied at the center lines of cutter and pulley respectively.
2. As a first approximation it is reasonable to assume that the shaft is of uniform cross section.
3. Cutter and pulley are dynamically balanced.
4. Use steel specified as 708M40 according to B.S. 970:1972 (SAE 4140), with the following minimum strength properties: $S_u = 700$ MPa; $S_y = 525$ MPa.

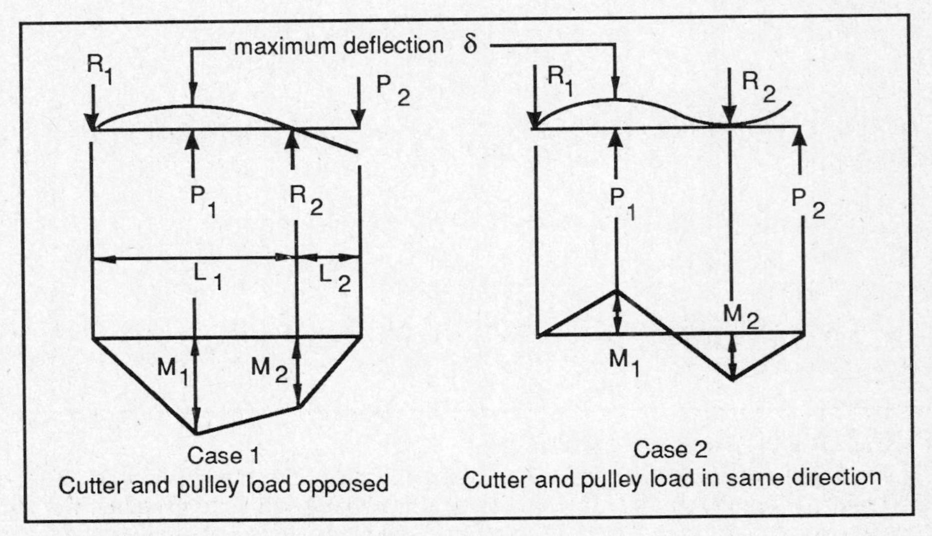

Figure 11.8 Bending moment and deflection diagrams

Solution

(a) Case 1 – Cutter and pulley loads opposed.

Power transmitted = 10 kW @ 4500 rpm

$$\text{Torque} = \frac{60 \times 10^{4}}{2\pi \times 4500} = 21.2 \text{ Nm}$$

forces at pulley and cutter :

$$\text{pulley force} = \frac{21.2}{0.055} = 383 \text{ N}$$

$$\text{cutter force} = \frac{21.2}{0.06} = 353 \text{ N}$$

Taking moments (refer figure 11.8) :

$R_1 = -277$ N ; $R_2 = 309$ N ; $M_1 = 58.3$ Nm ; $M_2 = 42.3$ Nm

Constraints

1. Deflection. Use $E = 210$ GPa

$$\delta = \frac{P_1 L_1^3}{48\ E\ I} + \frac{P_2 L_1^2 L_1}{16EI}$$

Maximum allowable deflection $\delta_{max} = 2.54 \times 10^{-4}$ m

$$\text{Thus } I \geq \frac{P_1 L_1^3}{48\ E\ \delta_{max}} + \frac{P_2 L_1^2 L_1}{16E\delta_{max}} = 1.038 \times 0^{-8} \text{ m}^4$$

yielding $d \geq 25.03$ mm; **d = 26 mm**

2. Stress. Using a stress raising factor of $K_{tf} = 2.0$ and a conservative design factor of safety $f = 4$, $S_y = 525$ MPa and $S'_e = 0.5S_u$; Marks Baumeister formula (equation 11.13) gives:

$$d^3 \geq 2.17 \times 4 \times \sqrt{.422\left(\frac{21.2}{525 \times 10^6}\right)^2 + \left(\frac{2.0 \times 58.3}{350 \times 10^6}\right)^2}$$

$d = 15.4$ mm

this means that deflection will govern this design.

(b) Case 2 – P_1 and P_2 act in same direction.

Taking moments (refer figure 11.8) :

$R_1 = 76$ N ; $R_2 = 662$ N ; $M_1 = 16$ Nm ; $M_2 = 42.3$ Nm (as for Case 1)

Clearly the previously checked case is worse.

11.7 Exercises in design for axial load, bending and torsion

11.7.1 Mobile crane

Figure 11.9 shows some salient features of a mobile crane. Attention is to be focussed on the hydraulic ram ('luffing ram') used to raise and lower the extendible arm of the crane.

The aims of this assignment are:

(a) to develop skill in the application of engineering analysis;
(b) to give practice in the design of engineering components to resist axial compression.

Data

It is required to design the 'luffing' rams of a 10 tonne (10^4 kg) mobile crane. There are two identical luffing rams situated on either side of the crane boom (refer to Figure 11.9). The purpose of these rams is to raise and lower the crane boom during its load cycle. The extending boom of the crane is designed for L_{max} = 16 m and the luffing rams have a maximum extended length of 4.2 m as shown.

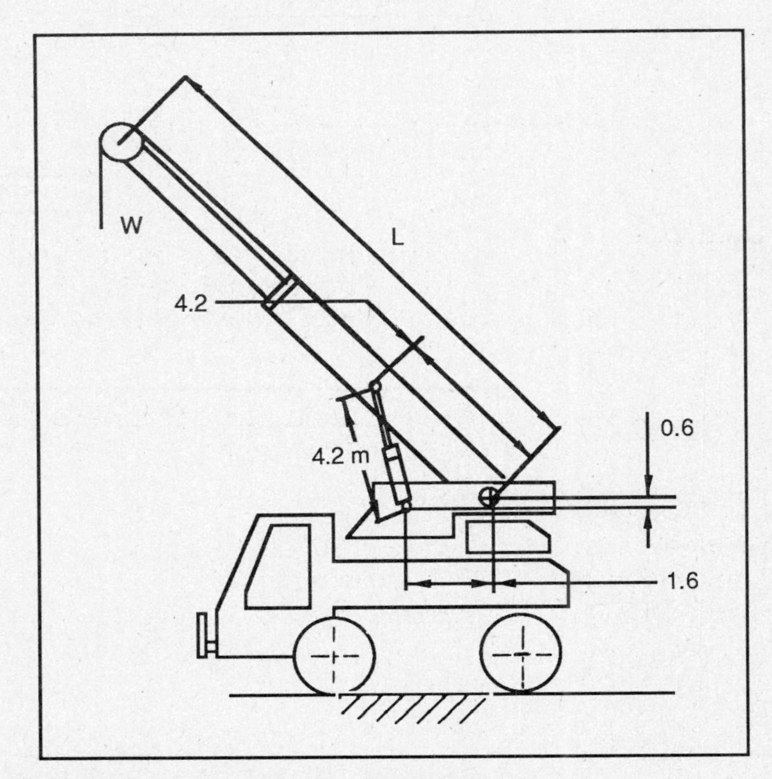

Figure 11.9 Travelling crane (all dimensions in meters)

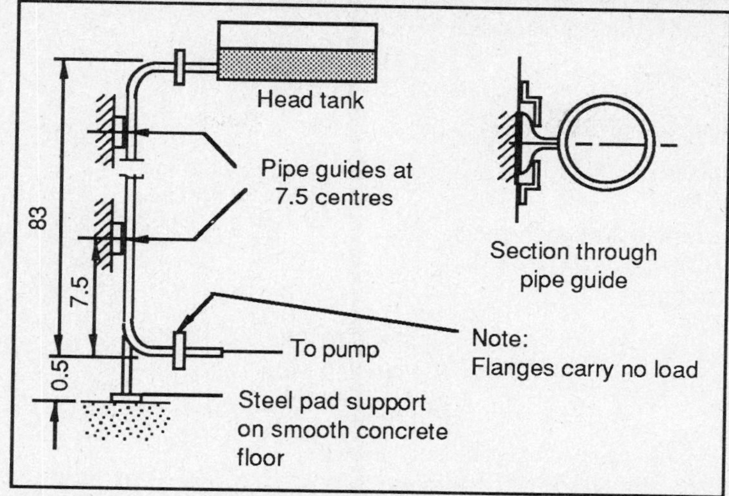

Figure 11.10 Cold water return pipe (dimensions in meters)

1. as a first approximation neglect the weight of the boom;
2. a design factor of safety of 2.5 will be used.

The design problem

1. Draw free body diagrams of the boom in several configurations to assess the load on the ram.
2. Determine the diameter of the ram.

Note: Neglect dynamic loads.

11.7.2 Cold water return pipe

A common method of supporting pipe-work in high-rise buildings is shown schematically in Figure 11.10. The pipe is shown supported by a short piece of pipe welded to the bottom bend, with a base plate resting on the service room floor. Apart from this support the pipe is self supporting; however, pipe guides are provided at regular vertical intervals along the pipe. These guides provide a degree of lateral restraint to the pipe. You as the responsible design engineer must establish the structural integrity of the pipe and support.

The design problem

1. (i) List the modes of failure of the pipe.
   (ii) Identify the sources of uncertainty affecting the selection of an appropriate factor of safety.
2. As the physical dimensions of the pipe have been decided on the basis of other design considerations, determine how safe the pipe is by estimating the factor of safety under the given structural conditions.
3. Estimate the load to be carried on the concrete floor. Comment on the proposed method of transmitting this load to the floor and any likely problems for the designer of the floor slab.

State clearly all design assumptions.

Notes

The following information refers to the steel pipe:

Inside diameter:	$d = 140$ mm
Outside diameter:	$D = 150$ mm
Second moment of area about a diameter:	$I_{xx} = 6 \times 10^{-6}$ m^4
Steel area of pipe:	$= \frac{\pi}{4}(D^2 - d^2) = 2.3 \times 10^{-3}$ m^2

$E = 210$ GPa
$S_y = 250$ MPa
$\sigma_B = 105$ MPa

where σ_B = allowable average contact pressure.
Density of steel = 8.1×10^3 kg/m^3.

Puzzle

This type of pipe arrangement is held together with flanges at regular intervals. The flanges often develop leaks early in the life of the pipe-work or during commissioning. In a city building just such a leakage developed in the first year of operation with a pipe slightly larger in diameter than that given in the problem. A large valve between the head tank and the first flange in the pipe (not shown in Figure 11.10) was shut off and a small valve was installed just downstream of this large valve and opened to atmosphere. The large valve just downstream of the pump (also not shown) at the bottom of the pipe was opened to clear the pipe of water. Unfortunately the time to clear the pipe seemed too long for the maintenance officer and, in his haste to empty the pipe, he decided to turn on the pump. The pipe in question was 250 mm outside diameter stainless steel with a 3 mm wall thickness.

1. Predict what happened to the pipe in the circumstances described above.
2. As the investigating engineer in the resulting court case, what information would you need to establish who should pay for any damages that may have occurred?

11.7.3 Scissor jack failure

A person using a 'scissor jack' from the tool-kit of a car with a mass of almost 2000 kg, has reported that it failed when the long central screw rod 'buckled'. Do you think this is possible? Give reasons for your answer.

Your task is to design the four 'arms' and central screw rod so that the jack shown in Figure 11.11 is suitable for inclusion in the tool-kit of an automobile weighing 2000 kg. Some decisions have already been taken, including the distances between the pin joints as shown, and that θ will range from 7° to 67° where the thread runs out. Steel to be used is AS1442-CS4140, with S_u 540 MPa and S_y 300 MPa.

Notes

1. Assume a uniform cross-section for each arm. For the arms, only calculate the minimum section properties, that is, cross-sectional area and/or second moment of area.

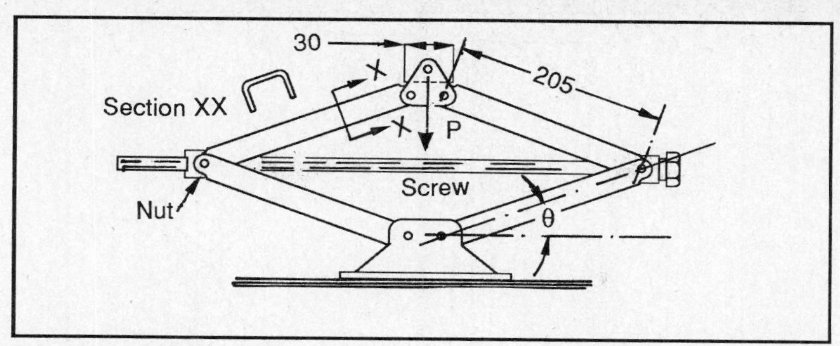

Figure 11.11 Scissor jack

2. Assume that the mean diameter $\left(\frac{\text{outside diameter + root diameter}}{2}\right)$ of a screw thread represents its effective diameter when supporting a load.
3. You will need to use free body diagrams to determine the worst loading situation(s).

Comment on any other information you would need to fully specify the arms and rod.

11.7.4 Playground maypole

A manufacturer of children's playground equipment has requested your services as a consultant in the design of a 'Maypole'.

The Maypole is a vertical steel pipe 2.5 m in height with a rotating cap at the top to which six 2 m length chains are attached, equispaced around the cap. At the other end of each chain is a ring where a child (or an adult occasionally) may hang on and swing around the Maypole. Figure 11.12 shows the Maypole schematically (not to scale).

In designing the steel pole, the manufacturer has made the assumption that the worst loading condition for the pole is a 150 kg adult swinging at a 3 m radius while running around the pole at 2 m/sec. These assumptions are based on the observation that an adult could achieve this speed running around a 'mock-up' Maypole at the manufacturer's premises.

1. What *modes of failure* should the steel pole be designed to resist?
2. *Estimate* an appropriate factor of safety for designing the pole to resist each mode of failure. Justify you estimate(s) as far as you are able.
3. *Determine* what size of steel pipe should be used to construct the pole. Commercially available sizes are shown in the table below.

Table 11.2 Steel pipe sections

Outside diameter (mm)	*Wall thickness (mm)*	*Area of cross-section* (m^2)	I_{xx} (m^4)
75	10	2.04×10^{-3}	1.10×10^{-6}
100	5	1.49×10^{-3}	1.69×10^{-6}
100	10	2.83×10^{-3}	2.90×10^{-6}
100	15	4.00×10^{-3}	3.73×10^{-6}
125	10	3.51×10^{-3}	6.01×10^{-6}

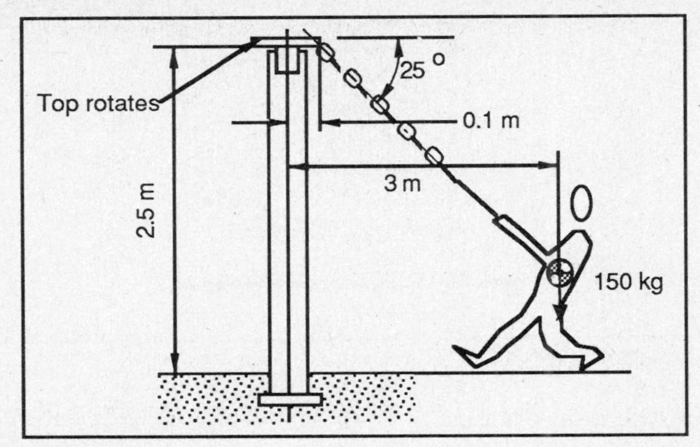

Figure 11.12 Maypole

4. *List* any important assumptions you have made in your design calculations. *Comment* on the loading assumptions. Are they realistic?

Notes

1. Pipe sections, other than those shown in the table above, are available, with outside diameters in multiples of 25 mm and wall thicknesses of 5, 10 and 15 mm.
2. The manufacturer of the steel pipe guarantees that the steel used has the following properties as determined by laboratory tests on specimens at his works:

Yield strength = 240 MPa
Ultimate tensile strength = 420 MPa
Young's modulus = 210 GPa

11.7.5 Industrial fuel tank

Figure 11.13 shows a schematic view of a small industrial fuel tank. Weight of tank when full is 10 kN. Containers of this type are to be made in large quantities with vertical supports welded to the corners as shown. To allow for specific geography of a given installation dimension A may vary from 400 mm to 3000 mm.

The information in Table 11.3 is available from Broken Hill Pty Ltd (a major manufacturer of rolled steel sections) for equal angles (this type of material will be used for vertical supports).

Table 11.3 Structural grade angle

B	*t*	*Mass*	*Section area*	*Radius of gyration*
mm	*mm*	*kg/m*	*mm*^{2}	(r_{xx} r_{yy}) *mm*
25	4.7	2.0	284	8.64
	3.0	1.07	155	7.37
32	6.0	2.54	361	10.2
	3.0	1.34	187	8.9
38	6.0	3.15	445	11.7
	3.0	1.61	239	10.4
50	6.0	4.29	607	15.0
	3.0	2.28	316	13.5

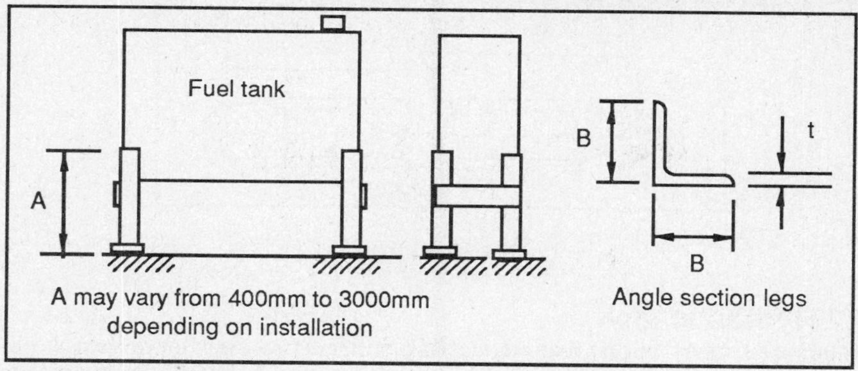

Figure 11.13 Industrial fuel tank

1. Specify the appropriate angle for the legs.
2. Specify the welds required at the corners. (Refer Chapter 13.)

Notes

1. Only one size is needed as all legs will be the same except for their length. However you should attempt to minimise the cost of the legs as far as possible.
2. Tanks will be free standing on floors of varying surface characteristics.
3. You may neglect side loads.
4. State clearly any assumptions made in arriving at a specification.

11.7.6 Tubular column design

Figure 11.14 shows a structural component which is to be designed to resist compressive load of 150 kN. It has been decided to use commercially available mild steel tubing with the following properties:

$$\text{Yield strength } S_y = 240 \text{ MPa}$$
$$\text{Young's modulus } E = 200 \text{ GPa}$$

Table 11.4 gives various section properties for commercially available mild steel tubing.

Table 11.4 Commercial mild steel tubing

Outside diameter (mm)	*Wall thickness (mm)*	*Area of cross-section (m^2)*	I_{xx} *(m^4)*	*Radius of gyration (mm)*
75	10	2.04×10^{-3}	1.10×10^{-6}	23.2
100	5	1.49×10^{-3}	1.69×10^{-6}	33.7
100	10	2.83×10^{-3}	2.90×10^{-6}	32.0
100	15	4.00×10^{-3}	3.73×10^{-6}	30.5
125	10	3.51×10^{-3}	6.01×10^{-6}	41.3

Determine the size of tube you would recommend for this application to give a factor of safety of 2.5.

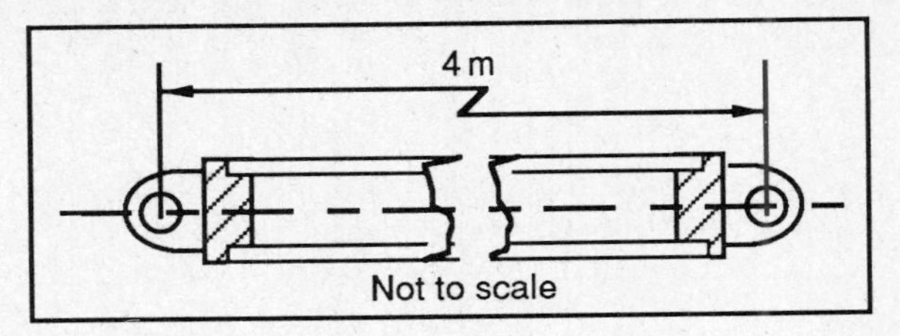

Figure 11.14 Tubular column

11.7.7 Hydraulic jack

The solution of many (most) real engineering problems requires more than the direct application of some theory. Often there is no theory which is exactly applicable, or the engineer does not have knowledge of such a theory. Often there is insufficient information (and there may be extraneous extra information), or there are insufficient constraints to yield an 'exact' solution (i.e. a number of 'solutions' will satisfy the stated requirements). It is also a feature of engineering design that an iterative procedure is often necessary before all of the requirements can be met.

This problem exhibits many of the above characteristics. You will need to make a number of initial decisions (assumptions) before you can proceed to use any theoretical analysis.

Problem

A common form of hydraulic hand 'jack' is shown in Figure 11.15. The piston/plunger is essentially a tube with a threaded insert at the top (see 11.15 (b)). The top extension screw is solid. The jack's rated lifting capacity is 2.25 tonnes, and the maximum hydraulic pressure is limited to 30 Mpa by a by-pass valve. The plunger can extend 130 mm, and the screw 50 mm. The fully retracted height of the jack is 187 mm.

The plunger and screw are to be made of hardened steel with S_y = 200 MPa. The depth of thread of the screw is 1.1 mm and its pitch is 3 mm.

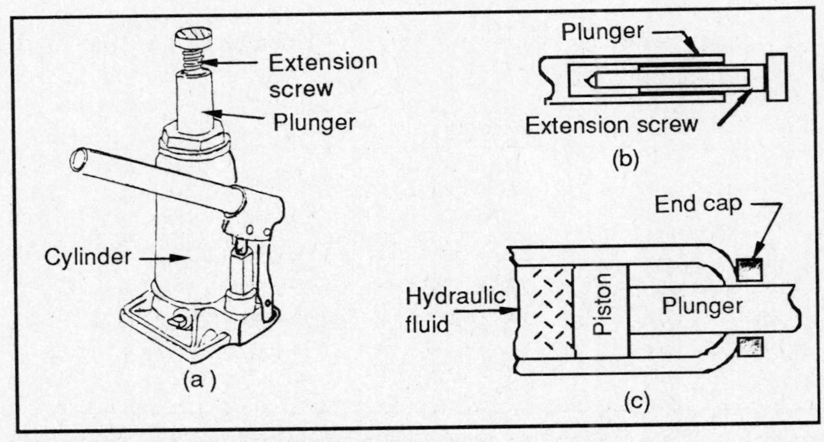

Figure 11.15 Components of a hydraulic jack

1. *Determine* the internal diameter of the hydraulic cylinder (to achieve the rated lifting capacity).
2. *Calculate* the size of the plunger (external diameter and thickness), and of the screw (major/maximum diameter).
3. *Choose* a suitable factor of safety. State all assumptions made, and justify these as well as you are able.

11.7.8 Centrifugal pump shaft

Figure 11.16 shows the layout of a motor-driven centrifugal pump. This is quite a common component of chemical and civil engineering systems. The pump operates at 2,900 rpm. (304 rads^{-1}) and absorbs 7.5 kW. The rotor weighs 178 N and is centrally placed between the bearings. The pump rotor weighs 51 N, and due to asymmetric pressure distribution around its periphery experiences a force of 56 N acting vertically downward.

The shaft is manufactured from a commercial grade of mild steel for which Young's modulus E is 210 GPa. The whole rotating assembly is dynamically balanced so that out-of-balance forces and moments are small and their effect on shaft deflections and stresses can be neglected.

The maximum permissible deflection of the motor and pump rotors at their center lines are respectively 0.10 mm and 0.18 mm.

1. Analyze the design of the shaft to determine whether it is satisfactory.
   Note that it would be reasonable to base calculations on a shaft of simpler geometry than that shown in Figure 11.16. A constant diameter of 45 mm could be assumed between the bearings, and a diameter of 28 mm to the right of the right-hand bearing, reducing to 22 mm at the bore of the pump rotor.
2. Comment on any important assumptions or simplifications made in your analysis.
3. The equations for elastic deflection of the shaft shown in Figure 11.16 are given below:

$$\delta_1 = \frac{P_1 l_1^3}{48\,E\,I_1} - \frac{P_2 l_1^2 l_2}{16\,E\,I_1}$$

$$\delta_2 = \frac{P_2 l_2^3}{3\,E\,I_2} - \frac{P_1 l_1^2 l_2}{16\,E\,I_1} + \frac{P_2 l_2^2 l_1}{3\,E\,I_1}$$

where: $$I_1 = \frac{\pi d_1^4}{64}\ ;\ I_2 = \frac{\pi d_2^4}{64}$$

11.7.9 Vertical hollow-shaft motor

Figure 11.17 shows a section through an electric motor. The motor delivers 7.5 kW at 1,470 rpm and was designed and manufactured in Melbourne. The motor is of a vertical, hollow shaft design used for powering irrigation pumps of the borehole type to provide spray or flood irrigation in country areas. The motor shaft is hollow to enable the pump shaft to pass through the motor and end in an adjusting nut at the top of the unit. The

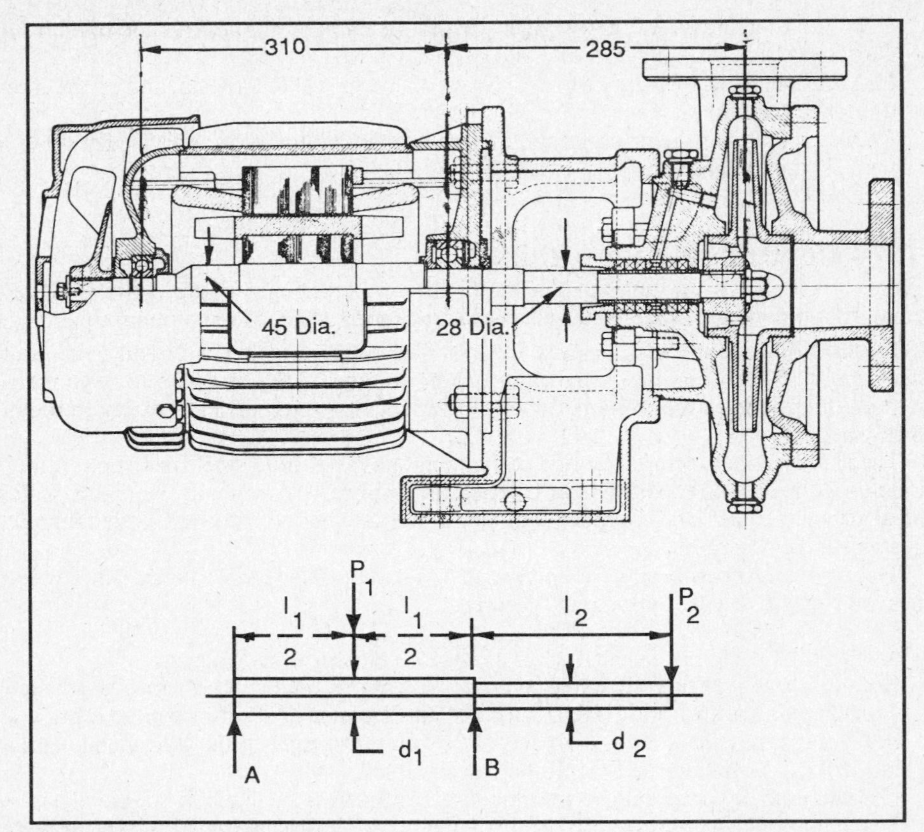

Figure 11.16 Centrifugal pump

operation of the pump produces a vertical downward force or thrust of 25 kN on the pump shaft. This is transmitted to the motor shaft and is ultimately carried by the lower ball bearing. In irrigation applications the power delivered by the motor is very nearly constant.

The motor shaft is specified as AS1442-CS4140 carbon-manganese steel with the following properties:

$$S_u = 540 \text{ MPa}, S_y = 300 \text{ MPa}, S_e = 270 \text{ MPa}$$

Its internal diameter is 25 mm; its external diameter at the upper bearing is 32 mm; and at the lower bearing is 40 mm.

1. (a) To what modes of failure is the motor shaft subject?
   (b) Where is the most highly stressed section of the shaft?
   (c) What are the stresses at this section? What is the maximum shear stress? Try to allow for any stress concentration effects.
   (d) Comment on the magnitude of the stresses you have calculated. Are they big enough to give any cause for worry?

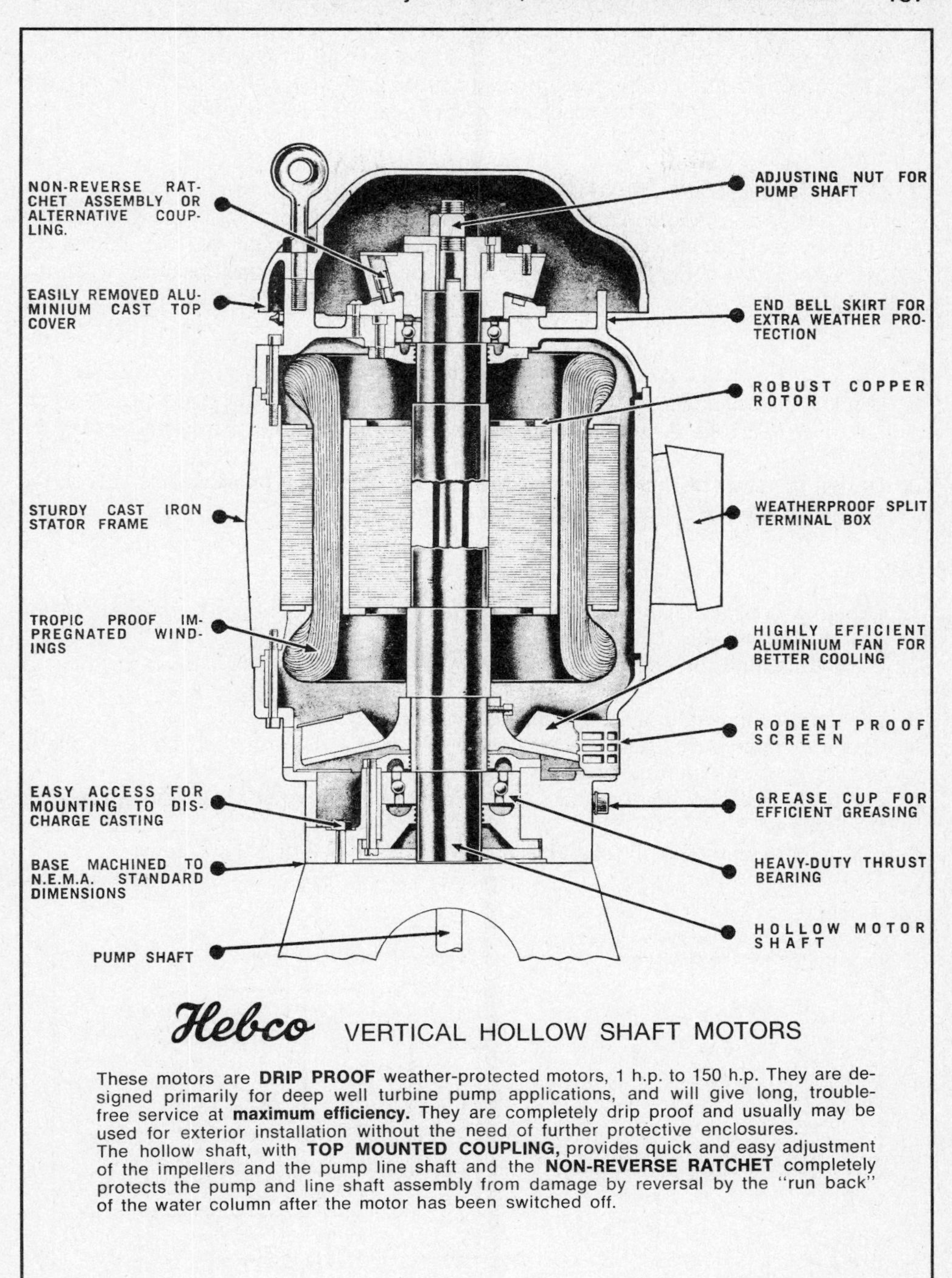

Figure 11.17 Hollow shaft electric motor

2. The air gap between the motor and stator is a critical dimension. What manufacturing errors or inaccuracies could adversely affect this dimension?
3. The motor manufacturer guarantees an efficiency of 89 percent. What are the sources of the losses? What problems do they pose for the designer?

11.7.10 Industrial centrifuge

Figure 11.18 is a schematic diagram of an industrial centrifuge. The drum is rotated at 500 rpm by the gear drive from a 10 kW motor. It is important that the axis of the centrifuge does not deflect more than 10 mm from the mean, central position at the top of the drum.

Problem

1. Discuss the argument in favour of both long and short bearing spans (the distance between bearings A and B) and suggest how a suitable compromise may be achieved.
2. Design the shaft of the centrifuge to satisfy the deflection requirements. Check the design for adequate strength, and adjust if necessary.

Notes

1. Design a constant diameter shaft (i.e. ignore the small changes in diameter at the gear and bearings).
2. Assume that all the deflection arises from bending in the shaft between bearings A and B.
3. The maximum weight of the drum and contents is 20 kN.
4. The maximum eccentricity of the center of mass of the drum and contents due to uneven distribution of the contents is 100 mm.
5. The position of the center of mass of the drums and its contents at maximum speed is 0.5 m above A.
6. The steel to be used in the shaft has the properties $S_u = 540$ MPa;

$$S_y = 278 \text{ MPa}; \quad E = 210 \text{ GPa}$$

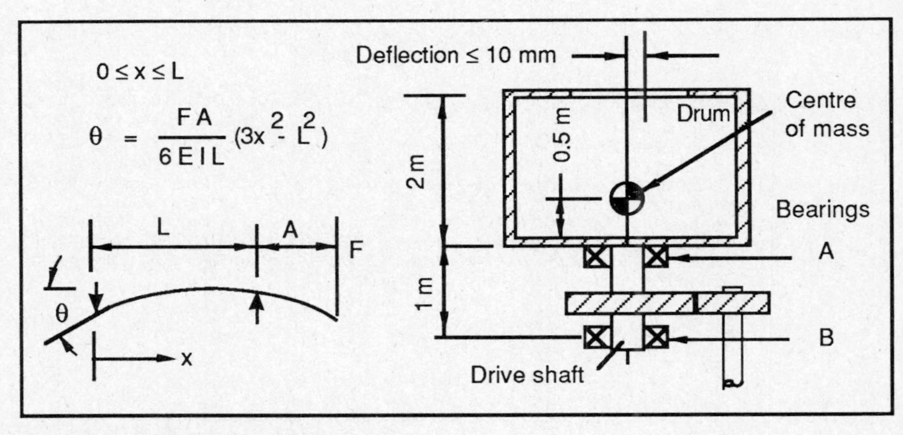

Figure 11.18 Industrial centrifuge

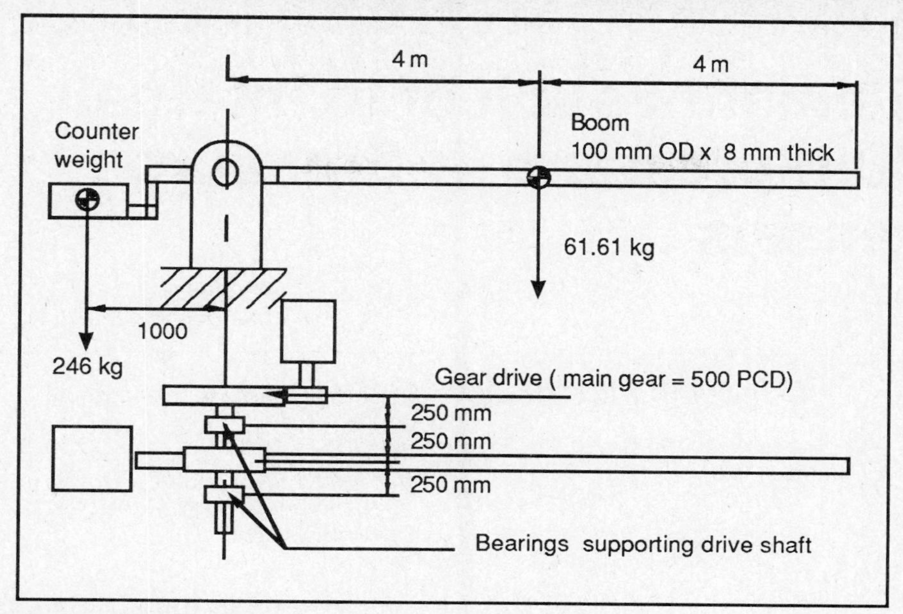

Figure 11.19 Boom gate

11.7.11 Boom gate

Many level crossings are serviced by boom gates. The operation of the gate is by an electric motor driving a main gear. The main gear is mounted on a drive shaft attached to the gate boom and the gate is raised or lowered by the direct stop-start action of the gear motor. Figure 11.19 shows a boom gate schematically.

Data

- Mass of gate = 61.6 kg.
- Mass of counterweight = 246 kg.
- Moment of inertia of gate and counterweight about the drive axle = 1546 kg m^2.
- Angular acceleration of the gate is assumed to be constant for 45° followed by constant deceleration to the full opening of 90° angular acceleration = $\pm\ 10^o/sec^2$.
- Material to be used is steel with specifications of AISI-SAE 6150 with S_u = 1500 MPa; S_y = 1000 MPa.

Design

The critical component of the design is the drive shaft. Sketch the forces on the shaft and sketch the bending moment diagram for both cases of positive and negative acceleration of the boom.

1. Show that the bending moment on the shaft in the plane of the boom is:

$$M_{max} = 501 \text{ Nm (approx.)}$$
$$M_{min} = 253 \text{ Nm (approx.)}$$

   and that the torque in the shaft is T = 247 Nm.
2. Specify the diameter of this shaft.

Chapter 12

Elements subject to internal fluid pressure — Pressure vessels

The man who smiles when things go wrong has found someone to blame it on.
Murphy

Concepts introduced — elements subject to triaxial stress conditions.

Methods presented — design of thin-walled cylindrical pressure vessels.

Application — containment vessel for an industrial centrifuge.

12.1 Introduction

We consider the design of *cylindrical pressure vessels* subject to internal fluid pressure, constructed from steel plates rolled to the correct curvature and welded together, and operating at moderate pressures and ambient or near-ambient temperatures high enough to exclude the possiblity of brittle fracture. This is an example of one type of fluid retaining element commonly used in engineering. Other examples are pipelines, hydroelectric penstocks, oil storage tanks, hydraulic and pneumatic control equipment. The principles of design to resist fluid pressure are similar in all these cases.

A comprehensive discussion of pressure vessel technology has been given by Nichols (1971). We will follow the relevant Australian Standard AS 1210, the 'Pressure Vessel Code', noting that it deals with unfired vessels (boilers are excluded) and that it covers non-ferrous materials of construction which are, however, of restricted applicability and outside the scope of this monograph. Other codes of practice for pressure vessel design and construction have been published by BSI, ASME, API, Lloyds Register of Shipping and VDI. The vessels have ellipsoidal, torispherical, or hemispherical ends, with plate thickness sufficiently small for thin-wall theory to apply, $t < 0.25D$ according to AS 1210, where t = minimum calculated plate thickness, D = inside diameter of shell in fully corroded condition.

AS 1210 classifies welded pressure vessels into four categories, Class 1 being the most special and Class 3 for general engineering service. The basis for the classification is indicated in Table 12.1. Figure 12.1 shows a typical welded vessel and illustrates the nature of the welded joints used in its construction.

The following data are available to the designer of a pressure vessel:

1. volume of fluid to be contained;
2. operating pressure and temperature as functions of time;

3. nature of the fluid, corrosiveness, and toxicity;
4. external loading (e.g wind load);
5. properties of available materials of construction;
6. location with respect to people and plant.

The designer must weigh up the risk, however remote, of the vessel failing and the consequences of such failure with respect to damage and injury sustained, costs of repair and lost production.

Table 12.1 Classification of pressure vessels

	CLASS 1	CLASS 2 2A	CLASS 2 2B	CLASS 3
Materials	Boiler plate, low to medium alloy steels			
	Structural grades of steel unacceptable			Structural grade mild-steel accepted
Design - weld joint efficiency η	0.95 to 1.0	0.8 to 0.85	0.7 to 0.75	0.5 to 0.65
Minimum shell thickness	Refer AS 1210			
Qualitycontrol procedures, inspection, tests and heat treatment	Very extensive	Extensive		Less extensive
Examples of applications	High pressure containment of toxic fluids, transportable vessels	Intermediate between classes (1) and (3)		General engineering, air receivers

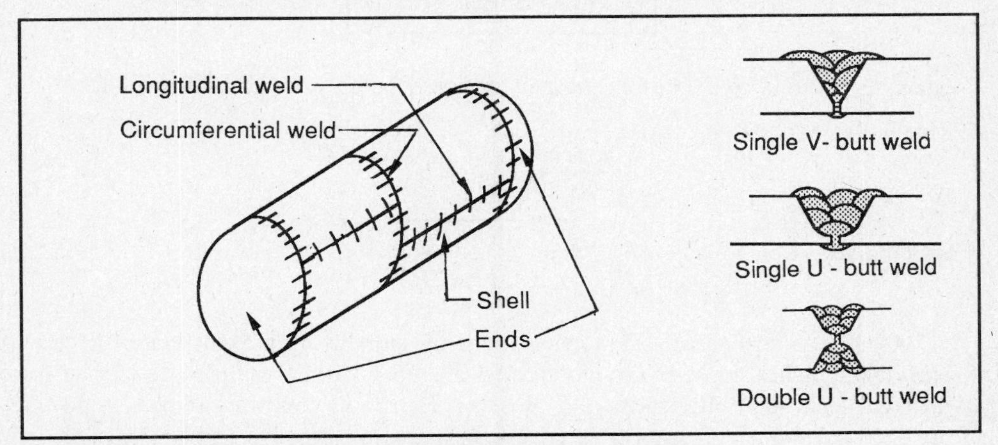

Figure 12.1 Typical welded joints used with pressure vessels

12.2 Design of the cylindrical shell

We consider the design of the cylindrical shell to resist internal pressure p, noting that in practice the designer may have to check the strength of the shell in resisting other loads, such as wind loads and loads due to the dead weight of the structure. The value of p used in design calculations will be chosen to exceed the maximum operating pressure by a suitable margin, perhaps by 5-10 percent in applications where safety valves are used. This is done in consideration of the possibility of pressure surges while the vessel is in service, as well as the need to prevent the unnecessary operation of pressure relief devices.

Mode of failure

Yielding, gross plastic deformation.

Mathematical model

$$\sigma_1 = \frac{pD}{2t} \quad \text{Circumferential stress}$$

$$\sigma_2 = \frac{pD}{4t} \quad \text{Longitudinal stress}$$

$$\sigma_3 = -\frac{p}{2} \quad \text{(approximation midway between O and } -p\text{)}$$

$$\boxed{\tau_{max} = \frac{(\sigma_1 - \sigma_3)}{2} = \frac{pD}{4t} + \frac{p}{4}} \qquad (12.1)$$

Theory of failure

The maximum shear stress theory predicts that a ductile material will yield when:

$$\tau_{max} = \frac{S_y}{2}$$

Factor of safety

$$\boxed{\tau_{max} \leq \tau_{all} = \frac{S_y}{2F_d}}$$

where $\frac{S_y}{F_d}$ would be the design stress in tension denoted f in AS 1210 and thus:

$$\frac{pD}{4t} + \frac{p}{4} < \frac{f}{2}$$

which gives:

$$\boxed{t \geq \frac{pD}{2f - p}} \qquad (12.2)$$

AS 1210 modifies the stress calculations by introducing a weld joint efficiency η and distinguishes between circumferential and longitudinal joints as they may have different weld joint efficiencies (η_L and η_C). Hence the following design inequalities are obtained.

Design inequalities

1. based on circumferential stress (longitudinal joints):

$$t \geq \frac{p\,D}{2f\,\eta_L - p} \tag{12.3}$$

2. based on longitudinal stress (circumferential joints):

$$t \geq \frac{p\,D}{4f\,\eta_C - p} \tag{12.4}$$

The minimum calculated thickness of the cylindrical shell must be the greater of (1) and (2). A suitable corrosion allowance c has to be added to the calculated value of shell thickness. For ordinary compressed air, steam, or water service $c = 1$ mm on each metal surface exposed to moisture.

12.3 Other design details

Ends

The design inequalities for determining the thickness of ellipsoidal, torospherical, and hemispherical ends are of the form (excluding corrosion allowance):

$$t \geq \frac{p\,DK}{2f\,\eta - 0.2p}$$

where K is a factor depending on the geometry of the vessel and is obtained from mechanics of solids theory. Recommended values of K are given in AS 1210.

Reinforcement of openings in vessel

The junction of a branch pipe with the shell of a vessel is a region of high stress, but the stress distribution is complicated and difficult to analyse. AS 1210 gives quick, approximate methods for determining what reinforcement, if any, is required at such junctions.

Supports

The designer is responsible for determining the way in which the vessel is to be supported. Usually a welded structure is used, and some typical examples are illustrated in AS 1210.

12.4 Other clauses in pressure vessel codes

Other clauses in AS 1210 deal with:

1. *Manufacture and workmanship*

   (a) preparation and cutting of plates;
   (b) tolerances;
   (c) welding procedures;
   (d) stress relief.

2. *Inspection and tests*

   (a) mechanical tests;
   (b) radiography;
   (c) pressure tests.

3. *Protective devices and fittings*

12.5 Notes from a designer's workbook

12.5.1 Industrial centrifuge

Figure 12.2 shows a cross-section of an industrial centrifuge used for separating dense material from an aqueous solution. The centrifuge is rotated at 500 rpm by the gear drive shown. At this speed, the more dense solid particles coat the inside of the drum to an average thickness of 150 mm. The water is then scooped out of the drum, the centrifuge is stopped and the wet particles are removed. Two lifting brackets are welded to the top of the drum as shown.

The design problem

1. What factor of safety (F_d) would you recommend be used in the design calculations for the drum?
2. If a factor of safety of 3.0 has been selected, determine the thickness of steel plate for the drum.
3. Design the lifting brackets for the shell.

   (a) Select a bolt size for the bracket.
   (b) Select a suitable thickness of material.
   (c) Design the welds for the bracket (refer Chapter 13).

Additional data

- Density of steel $\rho_s = 7{,}800$ kg m^{-3}.

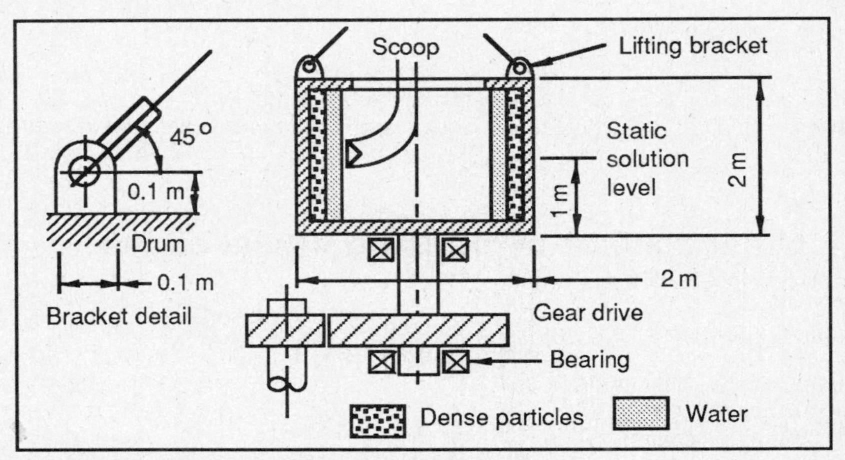

Figure 12.2 Industrial centrifuge (dimensions in mm)

- Yield strength of steel $S_y = 300$ MPa.
- Maximum allowable bearing (average contact) stress $S_b = 105$ MPa.
- The drum and welds will be 100% X-ray tested and coated with a special anti-corrosion paint.
- Density of wet particles in centrifuge $\rho_p = 1.6$ gm/cm^3.

Note

Pressure in a rotating fluid varies as the square of the radius of rotation by the relationship:

$$p = \rho\omega^2 r_2 (r_2 - r_1)$$

where:
- p = pressure at r_2.
- ρ = density of fluid.
- ω = speed of rotation.
- r_2 = outer radius of rotation.
- r_1 = inner radius of rotation.

Solution

(a) Safety factor

$$f_d = f_o S_1 S_2 S_3 S_4 S_5 l_1 l_2 l_3$$

Estimate

$f_o = 1.5$:	Failure is serious and dangerous.
$S_1 = 1.05$:	Steel is produced to a recognised standard.
$S_2 = 1.05$:	Welds fully inspected after manufacture.
$S_3 = 1.00$:	No corrosion.
$S_4 = 1.05$:	Possible stress increase through welding.
$S_5 = 1.10$:	Stresses at junctions may be higher than simple theoretical predictions.
$l_1 = 1.10$:	The drum may be overfilled in error.
$l_2 = 1.10$:	Dynamic force due to centrifugal forces and vibration is possible.
$l_3 = 1.10$:	The load may be off-center.

$$f_d = 1.5 \times (1.1)^4 \times (1.05)^3 = 2.54$$

(b) Drum thickness

Three sources of 'internal pressure' are:

(i) Water inside drum.

(ii) Particles inside drum.

(iii) Centrifugal stresses in steel drum.

(i) To find 'depth' of water in drum (refer figure 12.3):
Volume of fluid = $\pi \times 1^2 \times 1^2$ m^3 (half-filled drum)
Volume of cylinder of rotating fluid = $2\pi (r_2^2 - r_1^2)$
equating volumes and recalling that $r_2 - r_3$ is given as 0.15 m we get r_1 = 0.71 m and r_3 = 0.85 m.

Pressure due to water at water-solids interface

$$= 10^3 \times \left(\frac{2\pi}{60} \times 500\right)^2 \times 0.85 \times 0.14$$
$$= 326 \ kN/m^2$$

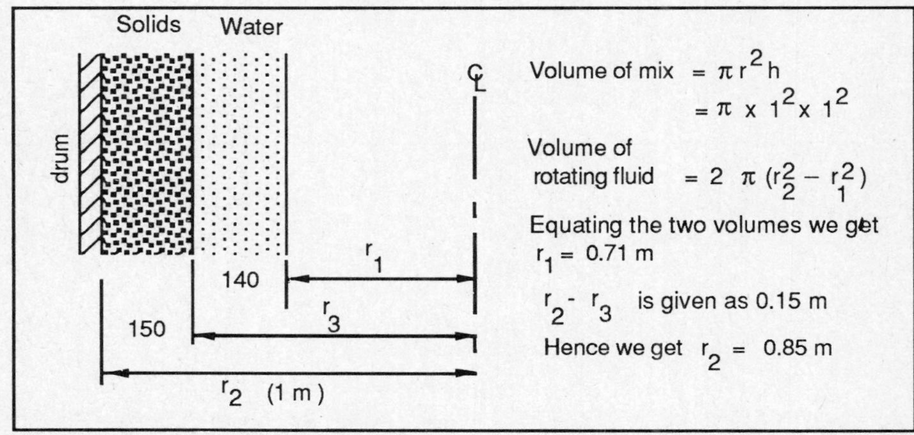

Figure 12.3 Geometry of rotating fluid

(ii) Pressure due to particles

$$= 1.6 \times 10^3 \times \left(\frac{2\pi}{60} \times 500\right)^2 \times 1 \times 0.15$$
$$= 658 \ kN/m^2$$

(iii) Centrifugal stress in drum
t = thickness of shell material in meters.
mass of 1 m^2 of shell

$$= 7.8 \times 10^{-3} \times 10^6 \, t$$
$$= 7.8 \times 10^3 \, t \ kg$$

Centripetal acceleration of drum material

$$= \omega^2 r$$
$$= \omega^2 r = \left(\frac{2\pi}{60} \times 500\right)^2 \times 1$$
$$= 2741 \ m/s^2$$

Hence, pressure on shell

$= 2741 \times 7.8 \times 10^3 \times t \ kN/m^2$

$= 21{,}400\ t\ kPa.$

Total internal 'pressure'

$= 326 + 658 + 21{,}400t\ kPa.$

(iv) Thickness calculation
Maximum shear stress theory of failure predicts ductile failure when

$$\tau_{max} = \frac{S_y}{2}$$

Equation (12.2) gives the shell wall thickness

$$t \geq \frac{p\breve{D}}{2f - p} \quad \text{where } f = \frac{S_y}{f_d}$$

In this case p is small compared to $2f$, so we approximate as follows:
(Assumption 1)

$$t \geq \frac{pD_i}{2f}$$

$$= \frac{(984 + 21{,}400t) \times 2 \times 10^3}{2 \times \frac{300}{3} \times 10^6}$$

solving for t we get

$t \geq 0.0125$

(yields p = 1,262 kPa compared to $2f$ = 200 Mpa
THEREFORE Assumption 1 is justified.)

Select t = 13 mm

(c) Lifting brackets (see Chapter 13 for method)
Approximate weight of drum = Volume of steel x 9.8 x density.
Assume drum is a complete cylinder with two ends from 13 mm steel

Area of sheet $\simeq 2 \times 2\pi + 2\pi = 6\pi\ m^2$

Weight of steel $= 6\pi \times 0.013 \times 7.8 \times 10^3 \times 9.8$

$= 18.73\ kN$

Assume weight of drum to be lifted is 20 kN
Force diagram of brackets (Figure 12.4)
shear force in bolts is approximately 14 kN

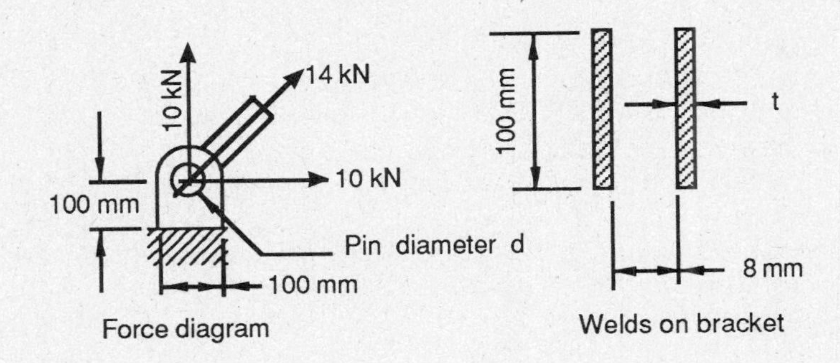

Figure 12.4 Forces and welds on lifting bracket

(i) bolt diameter d
Average shear stress in bolts (single shear)

$$= \frac{14000}{\pi d^2} \times 4. \qquad \frac{S_y}{2f_d} = 50 \text{ MPa}$$

Hence $$d^2 \geq \frac{14000 \times 4 \times 10^{-6}}{\pi \times 50} = 0.0189^2$$

$d = 20$ mm

(ii) Allowable contact pressure between pin and bracket is S_b

$$105\text{MPa} \geq \frac{\text{load}}{dT} \quad (T = \text{thickness of bracket})$$

Hence $$T \geq \frac{14000 \times 10^{-6}}{105 \times 0.020} = 0.067 : \text{Use 8 mm steel}$$

$T = 8$mm

(iii) Throat size of weld = t mm
Shear load = 10 kN

$$\text{Average shear stress} = \frac{10 \times 10^3}{2 \times t \times 10^{-3} \times 100 \times 10^{-3}}$$

$$= \frac{50.0}{t} \text{ MPa}$$

$$\text{Direct stress} = \frac{50.0}{t} \text{ MPa}$$

Bending moment = 10 x 0.10 = 1 KNm

Second moment of area for welds

$$\left(I_{zz} = \frac{BD^3}{12}\right)$$

$$= 2 \times \frac{1}{12} \times \frac{t}{1000} \times 0.10^3$$

Section modulus $Z = \frac{I}{y}$; $y_{max} = 0.050$

$$Z\ min = \frac{t}{6.0 \times 10^6 \times 0.050} = \frac{t}{3.0 \times 10^5}\ m^3$$

Hence bending stress $= \frac{M}{Z} = \frac{3.0 \times 105 \times 103}{t}$

$$= \frac{300}{t}\ MPa$$

Allowable stress in weld ≤ 96 Mpa

$$\sqrt{\left(\frac{50}{t}\right)^2 + \left(\frac{50}{t} + \frac{300}{t}\right)^2} \leq 96 = \frac{353}{t}$$

$$t \geq 3.63\ mm$$

weld size $\geq \sqrt{2}\,t = 5.2$ mm

Use 6mm weld

12.6 Exercises in design to resist internal pressure

12.6.1 Horizontal pressure vessel

Figure 12.5 shows a horizontal pressure vessel with hemispherical ends supported on two saddles 25 meters apart. The vessel will be used to store under a pressure of 1.55 MPa an organic liquid which is non-corrosive, non-toxic and non-inflammable. The density of the liquid is 1200 kg/m^3. The vessel will be of welded mild steel construction and it will have uniform wall thickness. The recommended material for this vessel is AS1442 — CS1020 mild steel with minimum strength properties of S_y = 270 MPa and S_u = 540 MPa.

Determine the wall thickness of the vessel.

Notes

1. Use the maximum shear stress theory of failure.
2. As far as you are able, justify the factor of safety you use.
3. Base your design on the stress in the cylindrical shell which is subject to both bending and internal pressure. (The stresses in the hemispherical ends will be lower than those in the cylindrical shell except at the junctions. In this case the junction stresses are very difficult to estimate because of the presence of the support saddles, but it is thought they will not be critical.)
4. Assume the vessel is simply supported. Ignore shear stresses due to bending; a

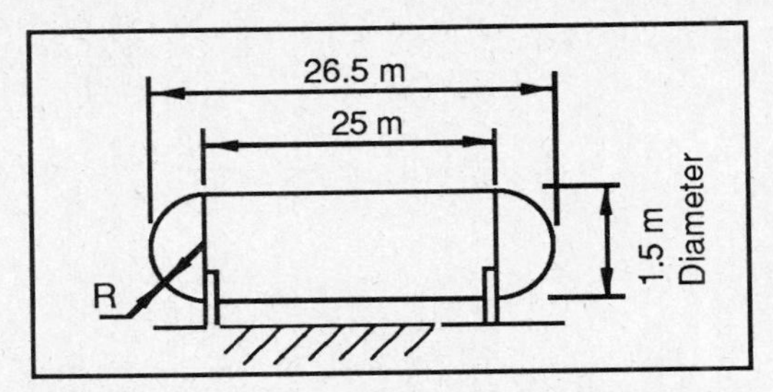

Figure 12.5 Horizontal pressure vessel (dimensions in meters, not to scale)

detailed analysis would show that they are zero in the most critically stressed regions of the cylindrical shell.

5. The distance of the center of gravity of a hemispherical shell of outer radius R and inner radius r from the flat diametral surface of the hemisphere is given by:

$$\frac{3(R^4 - r^4)}{8(R^3 - r^3)}$$

6. Some relevant equations are as follows (with usual notation):

- Longitudinal bending stress in an elastic beam is $\sigma_b = \frac{My}{I}$
- Longitudinal and circumferential stresses in a thin walled cylinder subject to an internal pressure p are $\frac{pD_i}{4t}$ and $\frac{pD_i}{2t}$ respectively.

7. It is your responsibility as the designer to make suitable approximations wherever possible in order to simplify unwieldy calculation.

12.6.2 Hydraulic ram

Figure 12.6 shows the layout of a double-acting hydraulic ram. This ram is intended for a load of 70 kN which may be applied in either tension or compression between the pinned-joint ends of the ram. The maximum oil pressure available to operate the ram is 4.75 MPa.

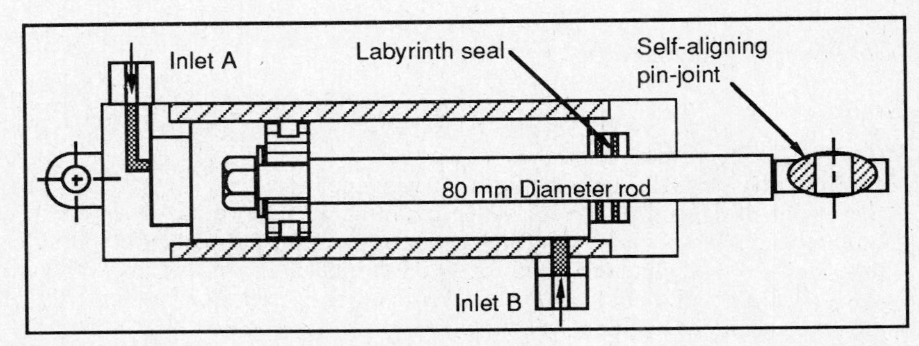

Figure 12.6 Double-acting hydraulic ram

Design problem

1. Draw free body diagrams of a ram and cylinder for both directions of loading.
2. Determine the diameter and wall thickness for the cylindrical shell of the hydraulic cylinder.
3. Specify the critical features of the pinned-joint end connection for the cylinder.
4. Prepare a dimensioned sketch of the cylinder.

Notes

1. Use a carbon manganese steel (AS1442 — CS4140) with minimum strength properties Su = 540 MPa, Sy = 300 MPa and bearing strength of Sb = 200 MPa.
2. The material used for the cylinder will be seamless steel tube with a weld joint efficiency of h = 0.95.
3. Neglect corrosion allowance.

Chapter 13

Joints for engineering components

The fact that an opinion has been widely held is no evidence whatsoever that it is not utterly absurd .

Bertrand Russell

Concepts introduced	joint stiffness; rigid and elastic behaviour of joints; weld behaviour.
Methods presented	design of bolted joints; estimation of torque to achieve a given bolt preload; failure modes of shear connectors; stresses in welded joints with parallel and transverse loads; design of welds supporting bending loads.
Applications	design of the flange bolts for a jack hammer.

13.1 Bolted joints

We consider engineering components connected by bolts. Our attention is focussed on connections where the bolts carry tensile loads (static or dynamic) due to load being applied to the engineering structure in which the joint is used. A typical bolted joint is illustrated in Figure 13.1.

Mode of failure

Failure of seal by loss of contact between gasket and the members being joined together.

Mathematical model

The usual assumptions made to construct a mathematical model to describe the behaviour of a bolted joint are:

1. the members being joined — a flange and cover in this case — are rigid; and
2. the gasket and bolt are elastic.

We wish to determine the tensile force in each bolt and the joint contact force when the joint is under full load.

The notation is as follows :

F = external force per bolt applied to the joint.

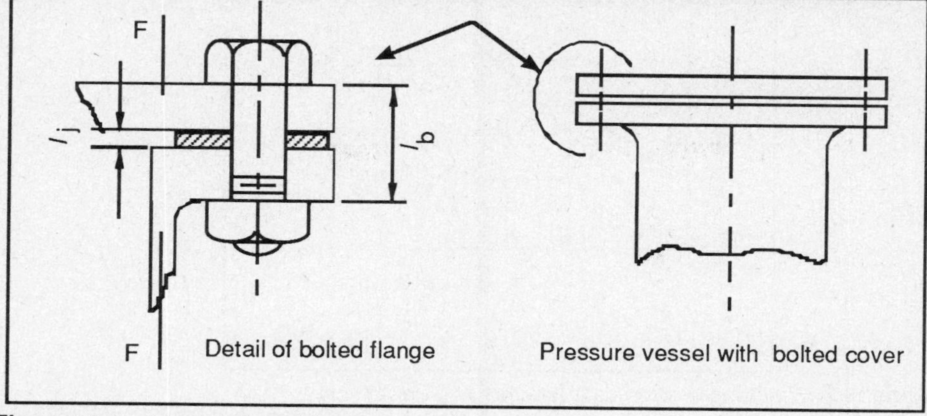

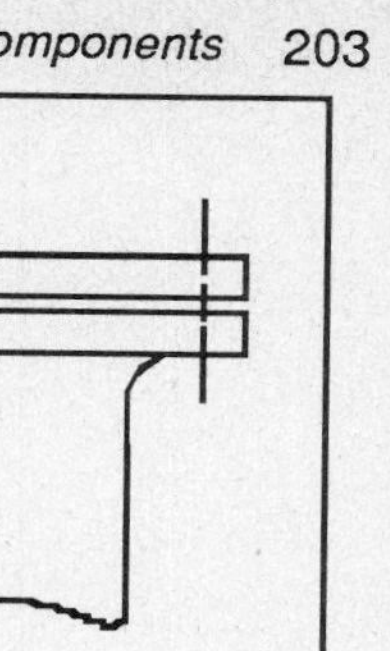

Figure 13.1 A typical bolted joint

F_j = joint contact force per bolt.
F_{ji} = initial value of F_j on assembly of joint (F = 0).
F_b = tensile force in bolt.
F_{bi} = initial value of F_b on assembly of joint (F = 0).
k_j = stiffness of joint $\left(\frac{A_j E_j}{l_j}\right)$.
k_b = stiffness of bolt $\left(\frac{A_b E_b}{l_b}\right)$

For equilibrium of the cover,

$$F_{bi} = F + F_j \tag{13.1}$$

and on assembly when F = 0

$$F_{bi} = F_{ji} \tag{13.2}$$

After application of the external load on the joint, the amount by which the bolt extends must be equal to the amount by which the gasket expands.

$$\frac{F_{bi} - F_{bi}}{k_b} = \frac{F_j - F_{ji}}{k_j} \tag{13.3}$$

Manipulation of equations (13.1), (13.2), and (13.3) yields:

$$F_b = F_{bi} + \left\{\frac{k_b}{k_b + k_j}\right\} F \tag{13.4}$$

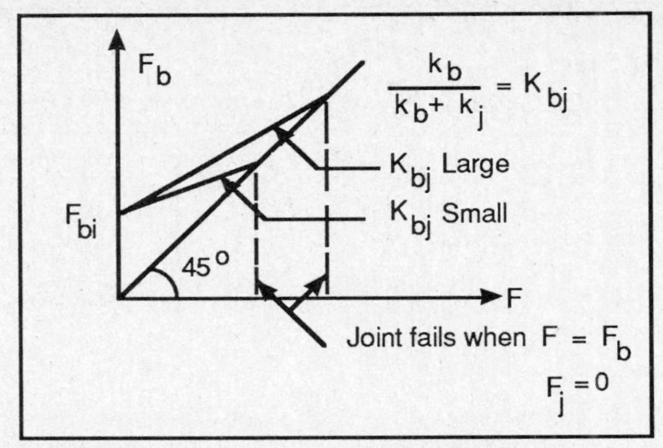

Figure 13.2 Bolt tensile force as a function of external force

$$F_j = F_{bi} - \left\{\frac{k_j}{k_b + k_j}\right\} F \tag{13.5}$$

The joint opens when Fj = 0. This occurs if and when:

$$F = \left\{\frac{k_b + k_j}{k_j}\right\} F_{bi} \tag{13.6}$$

It is instructive to plot:

1. bolt tensile force as a function of external force (Figure 13.2);
2. force as a function of deformation (Figure 13.3).

The second plot is useful because the graph can be modified to show the effects of plastic deformation of the bolts and also the effects of non-linear gaskets.

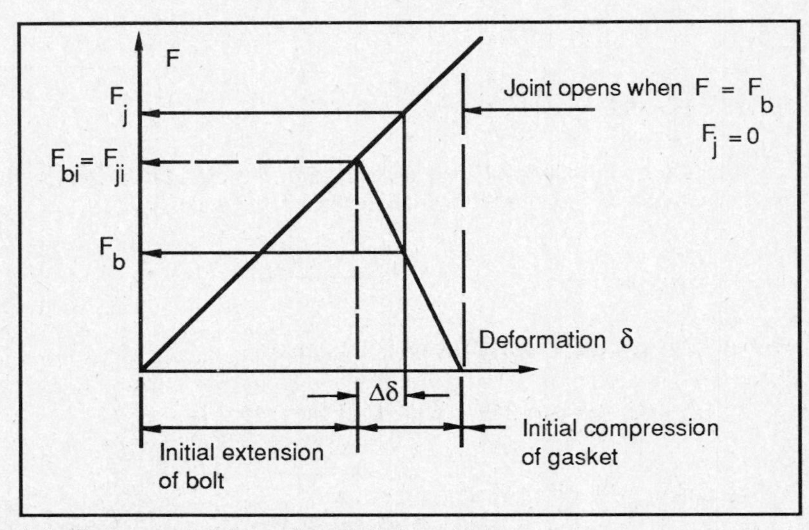

Figure 13.3 Forces and deformations in bolt and joint

Design inequality

F_{bi} must be greater than some prescribed minimum to ensure that the joint does not open.

$$F_{bi} > (F_{bi})_{min}$$

13.2 Bolting

A comprehensive discussion of the strength of screw threads is given in BS 3580, to which the interested reader is referred. The definitive articles on stresses in screw threads are presented separately by D. G. Sopwith and R. B. Heywood (Proceedings of I. Mech. E.). Both references are cited in the reference section of this volume and any thread designer should consult these excellent articles. In the design of threads it must always be borne in mind that the failure analysis should not only consider the nut or the bolt individually, but the combined action of these two elements.

Mode of failure

The most important mode of failure under steady load conditions is yielding of material in the body of the bolt.

Mathematical model

Bolt tensile force: $$F_b = F_{bi} + \left\{\frac{k_b}{k_b + k_j}\right\} F$$

Bolt tensile stress: $$\sigma_b = \frac{F_b}{A}$$

where A is the thread root area.

Factor of safety : $$\sigma_b \leq (\sigma_b)_{all} = \frac{S_y}{F_d}$$

where F_d allows for the effects of stress concentrations and any other uncertainties.

Design inequality

$$\frac{A}{F_b} > \frac{S_y}{F_d}$$

13.3 Comments on bolted joints

1. The assumptions made in constructing the mathematical model should be carefully noted. After all, flanged joints can twist and gaskets may deform under load in a non-linear manner. The possibility that the members being joined may themselves deform to a significant extent has been investigated by Samonov (1966).

2. Minimum bolt spacing is set by the space required for a spanner to grip the nut; maximum bolt spacing is limited by the possibility of fluid leaking past the gasket between the bolts.

3. It may be necessary to check stresses in the bolt during assembly, as well as the stresses in the thread (BS 3580). The torque T required to tighten a bolt and nut is related to the clamping force F_{bi} in the following way:

$$T = F_{bi}\left\{\frac{D}{2}\tan\lambda + \frac{D}{2}\,\frac{\mu_1}{\cos(x)} + \frac{d_o + d_i}{4}\mu_2\right\}$$

where:

d_i = diameter of inner edge of bearing area under the nut.
d_o = diameter of outer edge of bearing area under the nut.
D = pitch diameter of external thread.
x = flank angle of thread.
λ = lead angle of thread.
μ_1, μ_2 = coefficients of friction between mating threads and at bearing face of nut respectively.

The first term on RHS represents the torque absorbed in driving the mating thread helices over each other against the action of the axial load F_{bi} to which they are inclined. The second and third terms represent torques absorbed in overcoming friction between the threads and at the bearing face under the nut. In the case of an ordinary nut and bolt, a typical distribution of torque between the three terms would be, in the above order, 10 percent, 40 percent and 50 percent of the total. Friction conditions are thus of predominating importance, and, unless they are known accurately, there is no point in using an expression more complicated than the simple formula:

$$T = \frac{F_{bi}D'}{5}$$

where D' is the nominal thread diameter. For threads which are lubricated with at least a thin film of oil or grease, this formula is generally accurate to about 20 percent.

4. Fatigue loading: Suppose F fluctuates cyclically from 0 to some upper value F_{max}. Then the bolt tensile force fluctuates cyclically from F_{bi} to:

$$F_b = F_{bi} + \left\{\frac{k_b}{k_b + k_j}\right\}F_{max}$$

Therefore,

$$\frac{\text{amplitude of bolt force}}{\text{amplitude of applied force}} = \left\{\frac{k_b}{k_b + k_j}\right\}$$

In designing a bolted joint for cyclic loading it is desirable that the amplitude of fluctuation of the bolt force be kept low. It is thus desirable to reduce k_b (or $\frac{k_b}{k_j}$) and increase bolt flexibility, at the same time ensuring that the joint seal is maintained. This leads to special bolt designs of reduced shank diameter, used in conjunction with high F_{bi} and stiff packing. As shown top left in Figure 13.4 (a), a bolt of conventional design

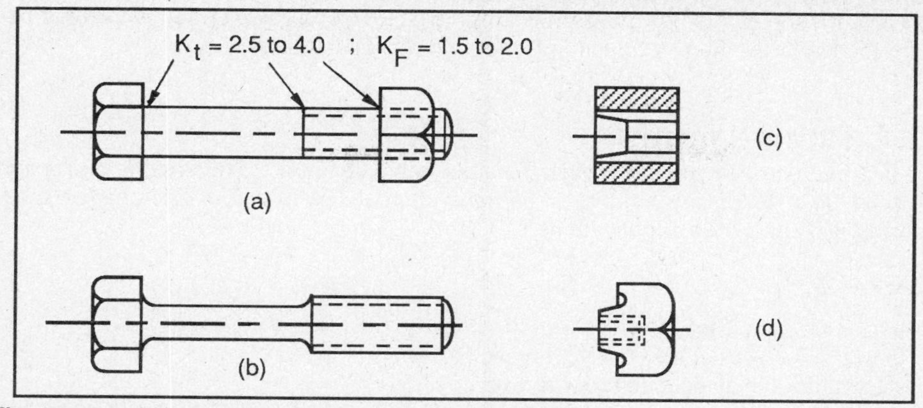

Figure 13.4 Designs of nuts and bolts for cyclic loading

has regions of high stress concentration so that a 15 percent reduction in shank diameter, as in Figure 13.4(b), has little or no effect on its fatigue life. Also shown on Figure 13.4 are alternative designs of nuts for increased fatigue life. In (c) the tapered part of the nut helps to distribute the applied load more evenly over the threads engaged with the bolt. In (d) the greater flexibility of the joint (higher k_j)has a beneficial effect on fatigue life.

The fatigue strength of bolts may be significantly improved by increasing the fillet radius at the root of the screw thread and by adopting a manufacturing process such as thread rolling which induces a surface compressive stress. On the other hand, the fatigue life of bolted joints is adversely affected by the presence of bending moments; it is important to ensure that as far as possible the bolts are subject to a purely axial load.

Useful information on the static and fatigue strengths of screwed connections is given in Vol. 5 of the Engineering Sciences Data Unit series on *Stress and Strength.*

5. Standards specification: The Society of Automotive Engineers (SAE) has designated a classification scheme for bolts, cap screws and studs. The standardized scheme allows for bolting to be specified for most applications by an appropriate SAE number. Table 13.1 gives some of the properties of bolts designated by the SAE scheme.

Table 13.1 SAE Bolt classification scheme

SAE grade	Minimum tensile strength (MPa)	Maximum hardness BHN	Material (steels)
1	379	207	Low carbon
2	379 — 475	207 — 241	Low carbon
3	689 — 758	269	Medium carbon
5	723 — 827	285 — 302	Medium carbon, heat treated
7	916	321	Medium carbon, alloy steel, heat treated
8	1034	352	Medium carbon, alloy steel, heat treated

Notes

1. As well as the above characteristics, SAE bolts are clearly marked on their heads with radial dashes equi-spaced by angle (e.g. grade 3 is designated by 2 marks at 180°; grade 8 is designated by 6 marks at 60°).

2. The variation in strength and hardness in the table is a function of bolt diameter, that is, the larger the diameter, the lower the strength and hardness.

13.4 Pinned joints

The distribution of stresses in material in a pin joint is complex. The ensuing discussion is based on a relatively simple mathematical model which is sufficient for most conventional engineering applications.

Modes of failure

1. tensile failure of rod;
2. tensile failure in net area of eye, Figure 13.5 (b);
3. shear failure in eye due to tear-out, Figure 13.5 (c);
4. tensile failure in net area of fork;
5. shear failure in fork due to tear-out;
6. compressive failure in eye due to excessive bearing pressure of pin;
7. compressive failure in fork due to excessive bearing pressure of pin;
8. shear of pin.

Mathematical model

σ_t is tensile stress, σ_c is compressive bearing stress, τ is shear stress.

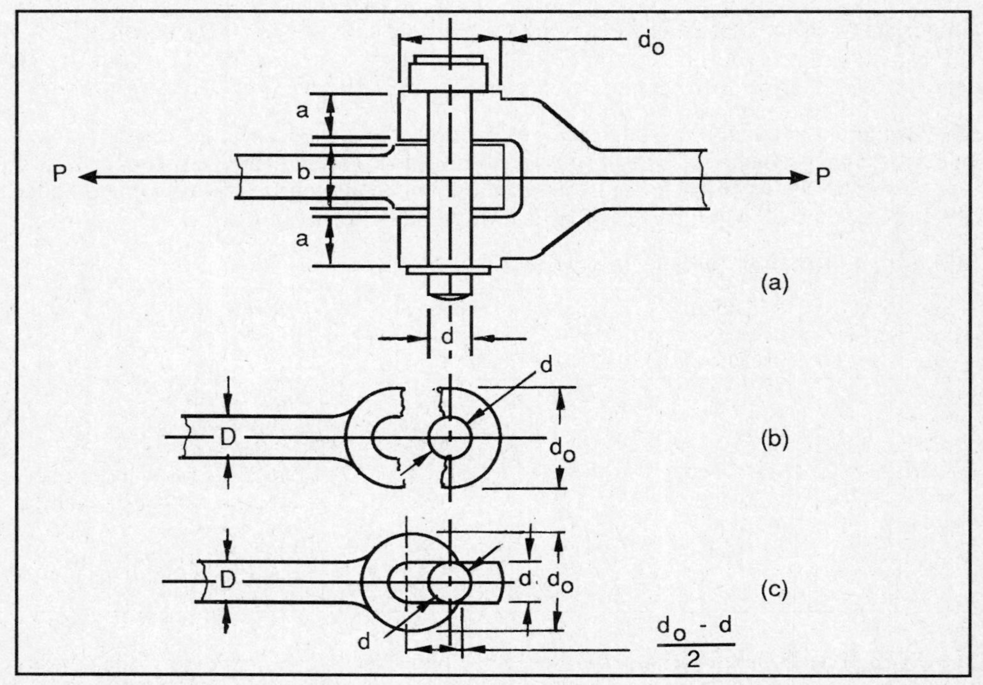

Figure 13.5 Failure modes in pin joints: (a) General view of joint (b) Tensile failure in the 'eye' (c) Shear failure in the 'eye'

1. $\sigma_t = \frac{4P}{\pi D^2}$

2. $\sigma_t = \frac{P}{(d_o - d)b}$

3. $\tau = \frac{P}{(d_o - d)b}$

4. $\sigma_t = \frac{P}{(d_o - d)2a}$

5. $\tau = \frac{P}{(d_o - d)2a}$

6. $\sigma_c = \frac{P}{d\,b}$

7. $\sigma_c = \frac{P}{2da}$

8. $\tau = \frac{P}{2} \Big/ \left(\frac{\pi d^2}{4}\right) = \frac{2P}{\pi d^2}$

Factors of safety

Factors of safety are implied in the selection of allowable stresses $(\sigma_t)_{all}$, $(\sigma_c)_{all}$, $(\tau)_{all}$.

Design inequalities

The stresses for conditions (1) to (8) above must satisfy:

$$\sigma_t \leq (\sigma_t)_{all}\,;\quad \sigma_c \leq (\sigma_c)_{all}\,;\quad \tau \leq (\tau)_{all}$$

13.5 Bolted joints — bolts in shear

A bolted joint can be designed to resist transverse shear loads in a plane perpendicular to the center-line of the bolts. The methods used are similar to those of Section 13.4 and hardly warrant separate discussion. The two most important modes of failure are:

1. shear of bolt; and
2. bearing, that is, excessive contact pressure between the bolt shank and the material around it.

If a number of bolts are used, the designer may have to decide on the basis of physical reasoning how the load is shared between them.

13.6 High tensile steel bolted joints

Bolted joints subject to transverse loads may be designed to act as friction joints for which the initial slip is so small that the bolts do not come into bearing on the sides of the holes. High strength steel bolts may be used in this manner when tightened in tension to at least 90 percent of the elastic limit in tension of the bolt material. When such bolted joints are properly tightened the members in contact develop a high frictional resistance so that slipping to bring the bolts into bearing does not occur under working loads. (Diameter of bolt holes is 1.5 to 2 mm larger than diameter of bolts.)

Tests have shown that when the calculated shear stress is reduced to 33 percent of the allowable bolt tensile stress any significant slip is eliminated. Design rules for this type of joint (e.g. those of the American Institute of Steel Construction) are based on this sort of experimental data; see, for example, AS 1511 on the 'Use of High Strength Bolts in Steel Structures'.

13.7 Introduction to welded joints

When two metals are welded together they coalesce to form common metallic crystals. For this to happen the layers of metallic oxide and absorbed gas on each metal surface have to be driven off. This is most frequently accomplished by heating, usually by electric arc or oxy-acetylene flame; the two metal parts fuse together locally. A weld is then surrounded by a 'heat-affected zone' of material, and proper control of the welding process must be exercised to ensure that unwanted metallurgical changes do not occur in this zone. This is indicated in Figure 13.6 which shows some typical welds.

The most common problem facing the designer of welded structures is the determination of the size of fillet welds. This is discussed below, but note that:

1. only one mode of failure is considered, namely the fracture of the joint;
2. factors of safety are not stated explicitly but are implied in the allowable weld stress chosen;
3. the stress in a fillet weld is considered as shear stress on the throat for any direction

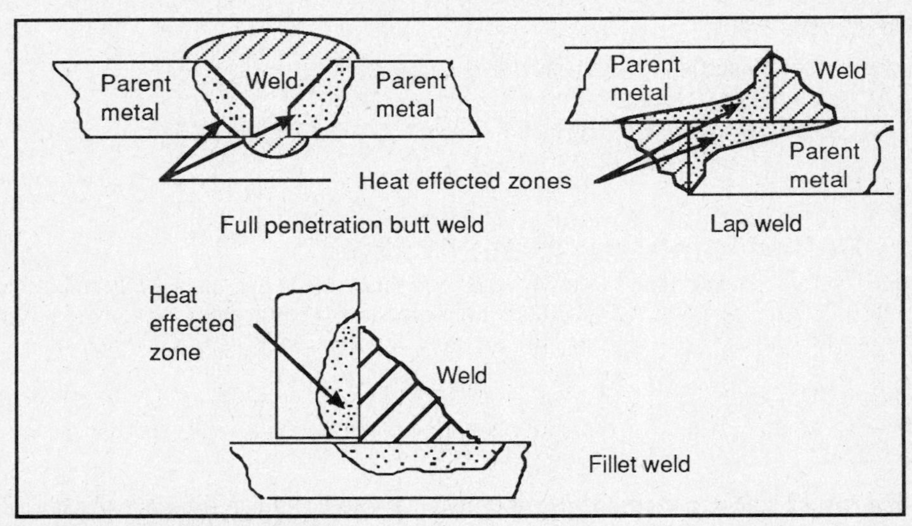

Figure 13.6 Some typical welds

of the applied load. In mild steel joints subject to steady loads the allowable shear stress is usually taken as 0.33 S_W, where S_W is the nominal tensile strength of the electrode used (the allowable shear stress is around 100 MPa, see AS 1250, clause 9.8.2). For fatigue loads this value is reduced, often by a substantial amount to allow for the adverse effects of stress concentrations, residual stresses, and weld cracks. (Gurney, 1968.)

13.7.1 Butt welds

Provided sufficient care is taken in manufacture, butt welded joints should be as strong as the parent metal. However, some design codes, such as AS 1210, require that the strength of the joint be reduced by a factor, the 'joint efficiency', to allow for the possibility of weld defects. The value used for joint efficiency depends on the care taken in manufacture and on the subsequent inspection and heat treatment of the weld.

13.7.2 Fillet welds — parallel and transverse loads

Parallel load

In this and subsequent discussion the following simplifying assumptions are made in order to construct a mathematical model:

1. The parent metal and the weld are homogeneous.
2. They both have the same properties.
3. Effects of stress concentrations are negligible.
4. Residual stresses are negligible.

Assumptions (3) and (4) are only really justified if special precautions are taken during manufacture.

Mathematical model

Symbols are defined in Figure 13.7 where:

F = load applied to welded joint
h = width of weld
l = length of weld
l_e = effective length of weld, usually taken as $(l - 2h)$
t = throat width of weld

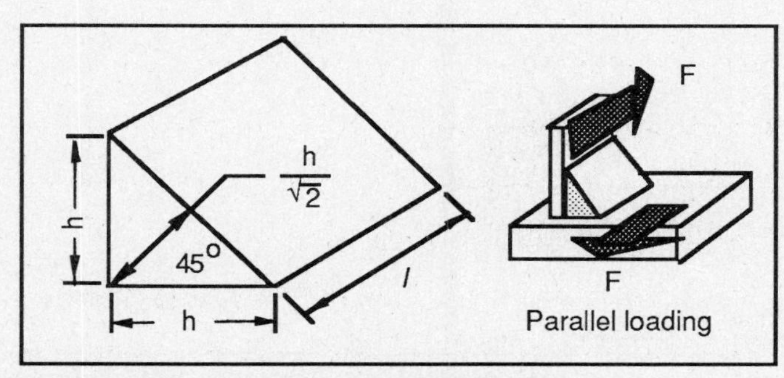

Figure 13.7 Nomenclature and shear load for fillet weld

Shear stress on throat $= \dfrac{F}{\dfrac{h}{\sqrt{2}} l_e} = \dfrac{\sqrt{2}F}{h(l - 2h)}$

Design inequality $\dfrac{\sqrt{2}\,F}{h(l - 2h)} \leq (\tau)_{all}$

Transverse load

Mathematical model

To construct the mathematical model it is assumed that the two members being joined are rigid. We investigate the throat stresses on the section defined by a plane T T at angle α, and do this by considering the equilibrium of the material shown shaded to the left of T T (refer to Figure 13.8).

The average shear stress on T T is:

$$\tau = \frac{F \sin \alpha}{x l_e} = \frac{F}{h l_e}\left[\sin^2 \alpha + \sin \alpha \cos \alpha\right]$$

This stress is a maximum when $\alpha = 67.5^o$ giving $\tau_{max} = 1.21 \dfrac{F}{h l_e}$

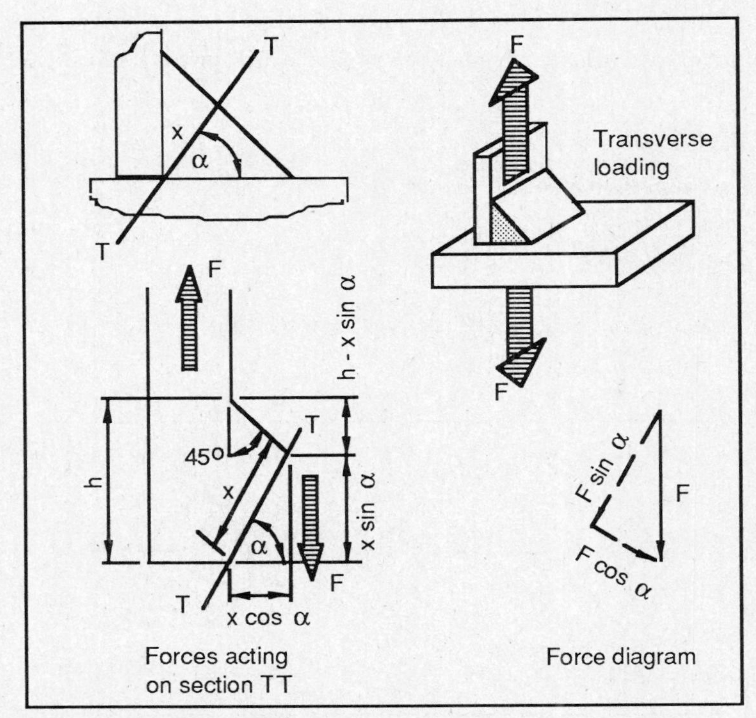

Figure 13.8 Transverse load on fillet weld

Design inequality

$$\frac{1.21\ F}{h(l - 2h)} \leq \tau_{all}$$

Parallel and transverse load

If part of the load is applied parallel and part transverse to the weld, the allowable parallel load should be used.

13.7.3 Fillet welds in bending

A simplified approach

Once again a simplified calculation is made of stresses on the weld throat. The following assumptions are made in bending calculations:

1. the bending stress in the weld is proportional to distance from the neutral axis;
2. any shear force is uniformly distributed over the length of the weld.

AS1250 specifies that where a fillet weld is subject to a combination of stresses, the equivalent stress must be less than an allowable value, that is:

$$\sqrt{f_n^2 + f_v^2} \leq \sqrt{3} \times 0.33\ S_w = 0.57\ S_w$$

where:

f_n = resultant direct stress normal to throat thickness.

f_v = vector sum of the shear stresses in the plane of the throat thickness.

S_w = nominal tensile strength of the electrode used in the weld.

Consider the application of the above rules to a simple example, namely the cantilever of constant circular cross-section shown in Figure 13.9.

$$\sqrt{\left(\frac{My}{I}\right)^2 + \left(\frac{P}{A}\right)^2} \leq \tau_{all}$$

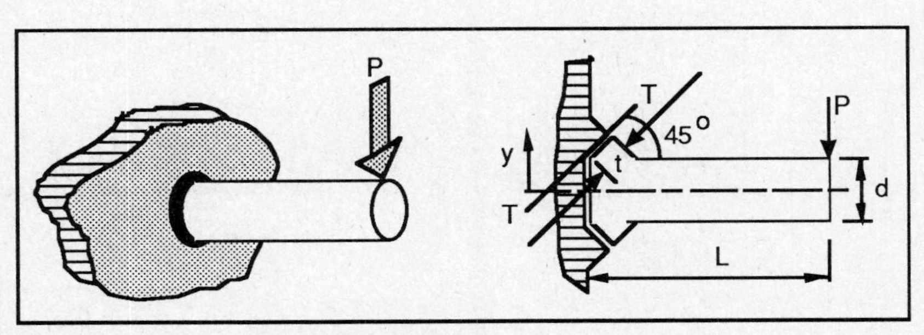

Figure 13.9 Cantilever of circular cross-section

where:

$$M = P L$$

$$y = \frac{d}{2}$$

$$A = \pi dt$$

$$I = \frac{\pi d^3 t}{8}$$

Hence, for a given τ_{all}, the throat width of the weld t can be calculated and its nominal size $h\ (= \sqrt{2}\ t)$ specified.

The American Welding Society (AWS) recommends a method of calculation which treats a fillet weld as a line and considers the resultant force per unit length the weld is capable of withstanding. While the AWS method appears different from the above, both procedures are similar in principle and give the same results.

13.7.4 Design of fillet welds based on allowable load per unit length of weld

Figure 13.10 illustrates fillet welds subject to parallel and transverse loads of magnitude F_p and F_t respectively. Other symbols are also defined.

Parallel load

Shear stress on weld throat is: $\tau = \frac{F_p}{tl_e} = \sqrt{2}\,\frac{F_p}{hl_e}$

The strength of the joint is expressed as the allowable load per unit length f_p where:

$$f_p = \left(\frac{F_p}{l_e}\right)_{all}$$

f_p may be related to the allowable shear stress on the weld throat τ_{all}:

$$f_p = \frac{h}{\sqrt{2}}\,\tau_{all}$$

For a given grade of steel, f_p depends on h.

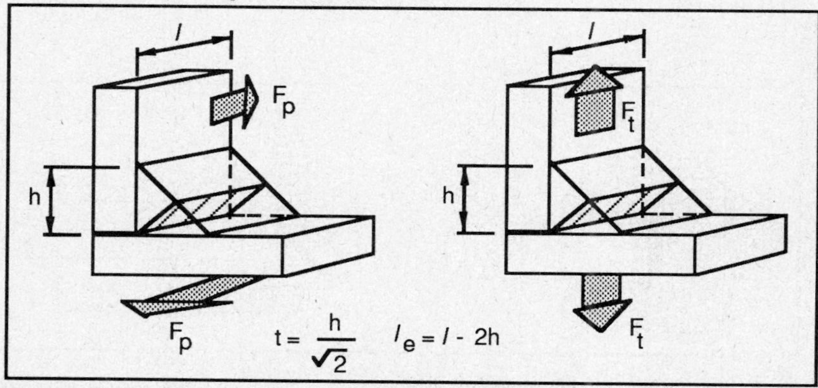

Fig 13.10 Fillet weld throat under parallel and transverse loading

Transverse and oblique loading

The strength of the joint under transverse load F_t is expressed as allowable load per unit length f_t:

$$f_t = \left(\frac{F_t}{l_e}\right)_{all}$$

For mild steel, or other grades of steel, f_t can be specified for different values of h. The implied relation with shear stress on weld throat is:

$$f_t = \frac{h\ \tau_{all}}{1.21}$$

If $\tau_{all} = 96$ MPa for mild steel and h is in millimeters, then $f_t = 80h$ kN per meter. This result is used for oblique loads, that is, loads which are a combination of transverse and parallel.

13.7.5 Fillet welds with bending and shear

As an example of the approach adopted, consider a cantilever of rectangular cross-section, fillet welded on its side faces to a vertical wall.

Under bending there is a horizontal force on each weld. Assume the force per unit length of weld is linearly distributed about the neutral axis of the cantilever as indicated in Figure 13.11. Let F_h be the horizontal force per unit length at the most highly stressed point in the weld — point A. Then the bending moment applied to the weld metal is PL, and this is resisted by a moment:

$$\frac{2F_h l_e}{4} \times \frac{2l_e}{3}$$

Hence F_h can be calculated for any known loading and geometry.

We further assume that the vertical shear force on the weld is uniformly distributed, so that the vertical force per unit length of weld is:

$$F_v = \frac{P}{2l_e}$$

The resultant force per unit length of weld at A is:

$$F_r = \sqrt{F_h{}^2 + F_v{}^2}$$

Knowing F_r, the weld is then sized as for oblique loading.

13.7.6 Fillet welds in torsion

Consider a cylindrical bar subject to an applied torque T, the bar being supported by a fillet weld around its circumference at one end (refer to Figure 13.12).

The allowable load per unit length of weld is f_p as in section 13.7.4. The resisting torque supplied by the weld is $f_p \times (\pi d) \times \left(\frac{d}{2}\right)$. The weld is sized so that:

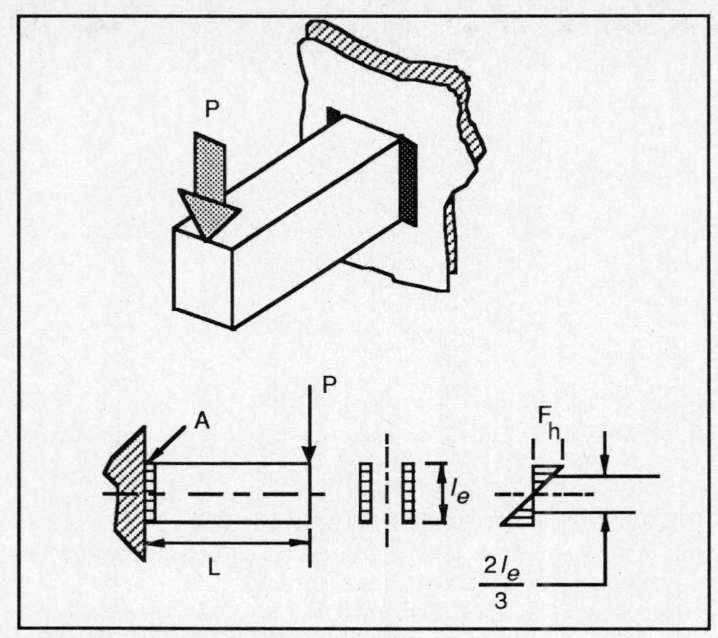

Figure 13.11 Fillet weld under bending and shear

$$f_p \left(\frac{\pi\, d^2}{2}\right) \geq T$$

13.8 Adhesives

A wide variety of adhesives is available for bonding surfaces of structural materials (Patrick, 1967). The strength of a glued joint is usually quoted in terms of the shear force per unit length it is capable of withstanding. Points to note are:

1. the importance of preparing the surfaces to be joined so that they are clean and have the required finish;
2. environmental effects — many adhesives are affected by extremes of temperature and humidity and by the presence of moisture. Lees (1984) discusses and explains the use of adhesives in engineering design.

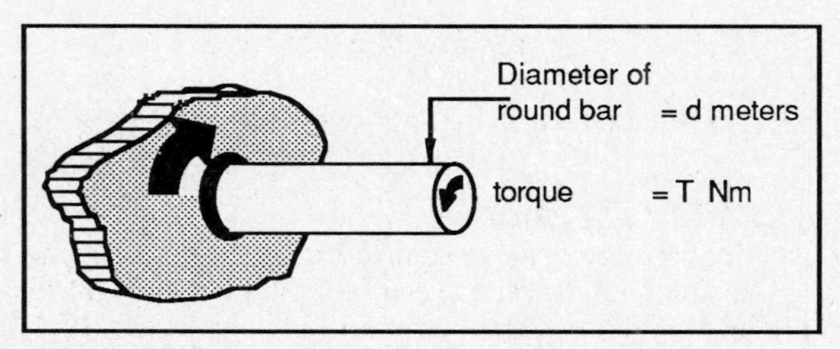

Figure 13.12 Fillet weld in torsion

13.9 Notes from a designer's workbook

13.9.1 Design of a bolted joint for a jack-hammer

The jack-hammer shown in Figure 13.13 is to be held together with either two or four threaded fasteners. The two tie rods will be used for this purpose, but these may be supplemented by two bolts which hold the upper cover to the cylinder.

It was shown in the jack-hammer problem of section 10.3.1 that it was necessary to ensure a high pre-tension in the tie rods to prevent them from experiencing compressive loads during the chisel rebound. Will this pre-tension be sufficient to maintain a gas-tight seal at the top of the cylinder?

If not, will two additional 8 mm bolts, shown in view A of Figure 13.13 and pretensioned to 1000 N, be sufficient?

Suggest a suitable re-design of these parts to conserve material and ensure a satisfactory seal.

Notes

1. The full gas pressure of 500 kPa may exist within the cylinder when the tie rods are at their minimum tension (shown earlier to be 600 N).
2. The tie rod and bolt material is steel with Young's modulus E = 210 GPa.
3. The gasket is a 1 mm thick, 2 mm wide ring of steel and asbestos with an elastic modulus E = 0.7 GPa.

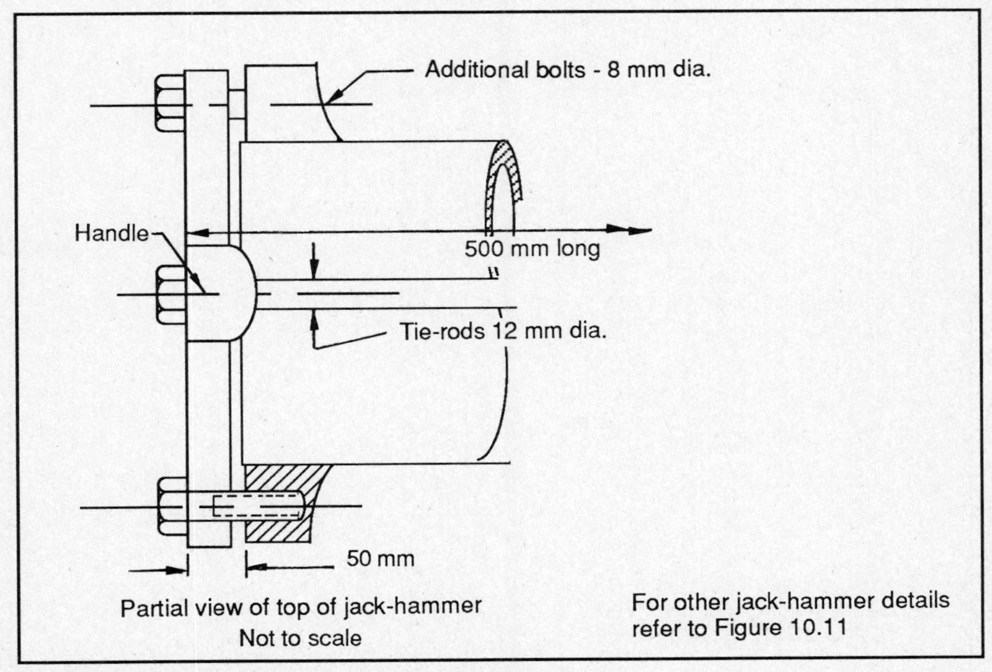

Figure 13.13 Schematic view of jack-hammer

Solution

(a) failure of the joint

(i) Maximum force on joint $= \text{pressure} \times \text{area}$

$$= 500{,}000 \times \frac{\pi}{4} \times (0.1)^2$$

$$= 3.93 \text{ kN}$$

(ii) Total force in tie rods (minimum) $= 2 \times 600$

Therefore $f_{ti} = 1.2 \text{ kN}$

(iii) Stiffness of each tie rod $= k_r = \dfrac{A_r E_r}{l_r}$

$$k_r = \frac{\pi}{4}\frac{(0.012)^2 \times 210 \times 10^9}{0.5} = 0.475 \times 10^8 \text{ Nm}^{-1}$$

Total stiffness of rods (2 rods) $= 0.95 \times 10^8 \text{ Nm}^{-1}$

(iv) Stiffness of the joint $= k_j = \dfrac{A_j E_j}{l_j}$

$$k_j = \frac{0.1 \times \pi \times .002 \times 7 \times 10^8}{.001} = 4.4 \times 10^8 \text{ Nm}^{-1}$$

(v) Joint opens when $f = \left(\dfrac{k_r + k_j}{k_j}\right) f_{ti}$

$$= \left(\frac{0.95 + 4.4}{4.4}\right) 1200$$

$$= 1.46 \text{ kN}$$

Since internal pressure forces may reach 3.93 kN, the joint will fail as designed.

(b) New bolts added

(i) For these bolts: $k_b = \dfrac{A_b E_b}{l_b}$

$$= \frac{\pi}{4}\frac{(0.008)^2 \times 210 \times 10^9}{0.050}$$

$$= 2.11 \times 10^8 \text{ Nm}^{-1}$$

Total stiffness for bolts and rods

$$k_{br} = (2 \times 2.11 + 0.95) \times 10^8$$

$$= 5.17 \times 10^8 \text{ Nm}^{-1}$$

(ii) Total bolt and tie rod force:

$$f_{bi} = 1200 + 2 \times 1000 = 3.2 \text{ kN}$$

(iii) Joint opens when

$$f = \left(\frac{k_b + k_j}{k_j}\right) f_{bi}$$

$$= \left(\frac{5.17 + 4.4}{4.4}\right) 3200$$

$$= 6.96 \text{ kN}$$

Joint with extra bolts will not fail.

(c) If the tie rods have a collar and mountings as shown below (figure 13.14), then k_r is increased because the length of the 'bolt' is reduced. The additional bolts are no longer necessary (this is shown below). If the tension in the rods still falls to 600 N, will the seal be maintained?

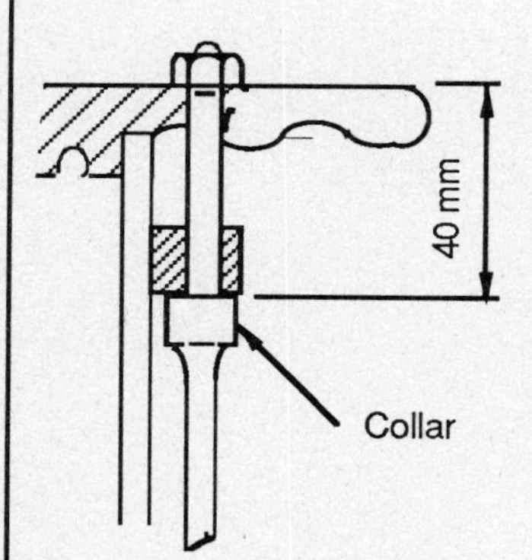

with this arrangement a pre-tension of 5.0 kN may be maintained in the upper part of the tie-rod, since this pre-load can be applied through the top nut. The tension in the lower part (to prevent buckling during the load cycle) will be applied to this section through the lower nut (refer to figure 13.13)

Figure 13.14 Proposed change in mounting of top tie-rod

Now

$$k_r = \frac{A_r E_r}{l_r}$$

$$= \frac{\pi}{4} \frac{(0.012)^2 \times 210 \times 10^9}{0.04}$$

$$= 5.9 \times 10^8 \ (\times 2) \text{ Nm}^{-1}$$

Hence joint opens when

$$f = \left(\frac{11.8 + 4.4}{4.4}\right) 1200 = 4.42 \text{ kN}$$

Since this load is more than the gas force, the seal is maintained. However, with the above arrangement, a pre-tension of 5000 N can be maintained in the upper part of the tie rod, since it will now be applied through the top nut. The tension in the lower part (to prevent buckling) will be formed by tightening the lower nuts (figure 13.13), and may fluctuate independently down to 600 N without affecting the joint at the top of the cylinder.

13.10 Exercises in the design of joints

13.10.1 Bolted joint for pressure vessel

Figure 13.15 illustrates the design of a bolted joint to retain a flat end cover on a cylinder subjected to an internal fluid pressure which fluctuates cyclically from zero (i.e. atmospheric pressure) to a pressure p. In the case shown, a full face gasket is used.

1. Other useful symbols are n, the number of bolts; A_r , the thread root area of bolts; P_b, the chordal bolt pitch.
2. Thicknesses of the cover and flange (dimensions t_c and t_f in Figure 13.15) are determined by strength considerations to keep stresses and deflections within acceptable limits.
3. Grade 3 bolts are to be used with a K_t of 1.5, a K_F of 2.2, and K_s of 1.1.

Design problems

1. For the particular application, the following information is available from design decisions already taken:

Internal fluid pressure (p)	=	5 MPa
Internal diameter of cylinder (D_i)	=	125 mm
Thickness of cylinder wall (t)	=	6.5 mm
Number (n) and size of bolts	=	8 - M10
Thread depth	=	0.8 mm
Ratio of gasket stiffness to bolt stiffness $\frac{k_g}{k_b}$	=	8.0
Joint thickness (l_j)	=	3.8 mm
Bolt material is alloy steel with yield strength (S_{yb})	=	380 MPa
Tensile strength (S_{ub})	=	620 MPa
Endurance limit (S_{eb})	=	310 MPa

(a) Determine the initial tightening load that must be applied to each bolt to prevent the separation of the joint.

(b) Determine whether the bolts are safely loaded, given that they are initially loaded by an amount 25 percent greater than that calculated in (a), and that a design factor of safety must not be less than 3 to account for uncertainties.

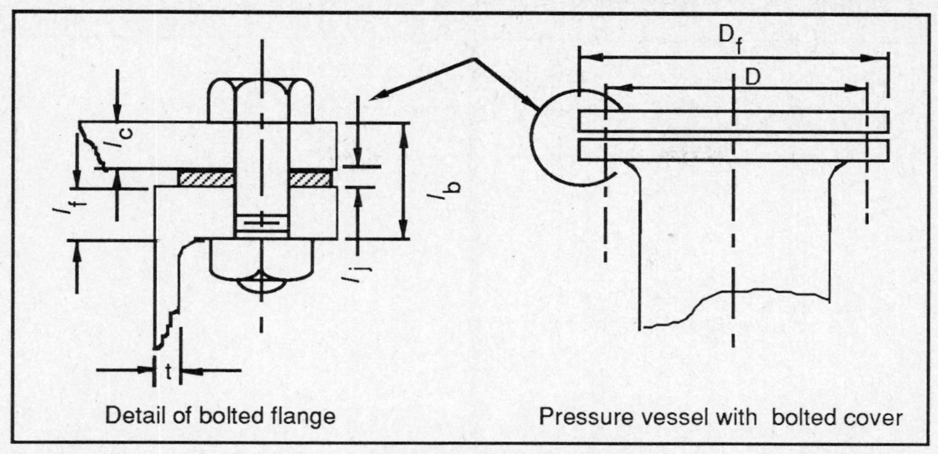

Figure 13.15 Pressure vessel with bolted cover (bolt diameter is d; vessel internal diameter is D_i)

2. Consider the general problem of designing a joint where the following are given or have been calculated previously: p, D_i, t, t_c, t_f.
   (a) List the quantities (controllable design variables) to be determined by the designer.
   (b) List the ways in which the joint may fail and state the corresponding inequalities which must be satisfied.
   (c) Write down the equations which constitute the mathematical model on which the design calculations are based.
   (d) Describe the strategy you would use to design the joint. Comment on any interesting features.

13.10.2 Cylinder-head studs

The cylinder head of a single cylinder internal combustion engine is bolted to the cylinder block by six M12 studs equally spaced around the bore. The nuts are tightened to give a tensile stress of 80 MPa in the unthreaded, 9.5 mm diameter, portion of the studs. The bore of the cylinder is 75 mm diameter and an annular copper gasket, 75 mm inside diameter, 125 mm outside diameter and 1.5 mm thickness is used.

1. *Determine*:

(a) the tensile stress in each stud when the cylinder pressure is p = 5000 kPa;
(b) the cylinder pressure when the joint opens.

Note
The following additional information has been obtained and decisions made in order to construct the mathematical model of this problem:

1. Modulus of elasticity of copper = 105 GPa.
2. Modulus of elasticity of cast steel = 210 GPa.
3. The cylinder head is a steel casting and must be regarded as compressible. The equivalent head may be considered as an annular cylinder 12 mm inside diameter, 25 mm outside diameter and 50 mm long at each stud.

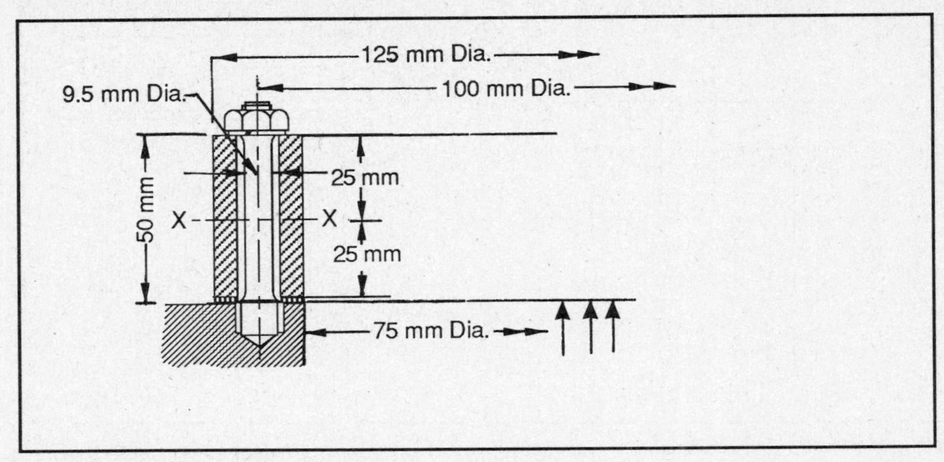

Figure 13.16 Cylinder-head studs

4. The applied force F due to the internal pressure is distributed through the cylinder head and is finally resisted by the contact force at the underside of the nut (see Figure 13.16). For these calculations it will be assumed that this force is transmitted through the cylinder head in such a way as to be equivalent to a force halfway up the cylinder head. In other words, it is assumed that the upper half of the cylinder head is compressed by this force while the bottom half is unaffected by it (see Trans. Inst. Eng. Aust., Vol. MC2, No. 2, Nov. 1966, for a more detailed discussion of this problem).
5. The properties of the stud material allowing for the effects of elevated temperatures and stress concentrations are estimated to be:

Endurance limit	=	80 MPa.
Yield strength	=	140 Mpa.
Ultimate tensile strength	=	210 MPa.

2. *Investigate* whether it is possible for the studs to fail before the joint opens. Modes of failure to be considered are:

(a) yielding; and
(b) fatigue fracture.

3. *Comment* on the use of an A-M diagram to investigate the possibility of a fatigue failure in this application.

13.10.3 Hydraulic actuator

Figure 13.17 is a schematic sketch of a linear hydraulic actuator. The working pressure of the actuator is 4.2 MPa, to provide approximately 32 kN of maximum load with a fundamental frequency of 10 Hz. You may assume the load F to behave sinusoidally according to F = 32 sin ωt kN. There are 4 SAE grade 3 bolts holding the cylinder together (S_u = 700 MPa).

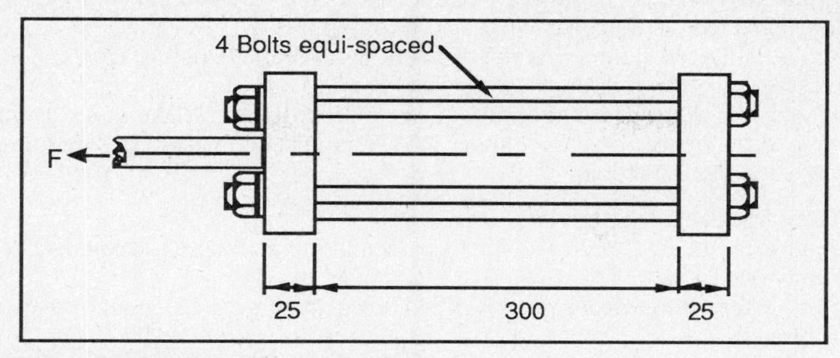

Figure 13.17 Hydraulic actuator (not to scale)
Note: 1. All dimensions are in millimeters.
2. Operating pressure is 4.2 MPa.

Determine the design details of the bolts holding the actuator cylinder together during operation. The following steps are suggested:

(1) Assume a value for the bolt diameter.
(2) Calculate the pre-load necessary to prevent leakage during operation.
(3) Find mean and fluctuating stresses in the bolts, and use an A-M diagram to determine safe operating conditions for the bolts assuming infinite life.
(4) See Samonov, 1966, for a detailed discussion of this problem.

Note
The following assumptions may be applied:

1. The cylinder ends are completely rigid.
2. The cylinder is mild steel with S_u = 400 Mpa, and it is assumed that during pre-loading the cylinder is uniformly compressed.
3. The seal providing a flexible joint between the cylinder and its ends is made of aluminum with a modulus of elasticity of 70 GPa.

13.10.4 Flanged coupling

Figure 13.18 shows a rigid, flanged coupling. This type of coupling is used to transmit power from one shaft to another adjacent to it, the two shafts being accurately aligned. Torque is transmitted from one shaft via a key to the corresponding half coupling, then via a number of accurately machined and fitted bolts to the other half coupling and then to the other shaft. Some torque may be transmitted by friction between the two half couplings if the coupling bolts are sufficiently tight. However, it is difficult to estimate this friction torque, and for this reason you may neglect it in your calculations.

It is required to design a rigid, flanged coupling for which the following information is available as a result of the previous design decisions that have been taken:

- 50 kW is to be transmitted at 1,470 rpm.
- The shaft diameter d is 40 mm.
- The shaft and keys are of mild steel with a minimum tensile strength of 420 MPa and yield strength 250 MPa.
- The cast iron half couplings have a minimum tensile strength of 150 MPa.
- The keys are to be of square cross-section.

1. *Explain* why theories of failure and factors of safety are used in engineering design. (Suggested length of answer 100 to 200 words.)
2. *List* the modes of failure that the different parts of the coupling should be designed to resist.
3. *Estimate* the allowable stresses for each mode of failure. Make quick estimates, do *not* go to the trouble of estimating or calculating explicit factors of safety.
4. *Determine* the following:
   (a) key width and length;
   (b) number and diameter of coupling bolts and the pitch circle diameter D on which they lie;
   (c) any other dimensions which affect the capacity of the coupling to transmit torque.

13.10.5 Welded bracket

The bracket shown in Figure 13.19 is to support 50 kN applied in the range shown $\pm 60^o$ relative to the vertical. The load is applied from a lifting device located below the bracket and used for moving equipment on a production line. The brackets are to be made from structural grade steel (AS 1442-CS4142) and the bolts are of high tensile steel, SAE grade 8 bolts, with a yield strength of 1000 MPa. Bearing strength for the bracket may be taken as 400 MPa.

Specify the critical dimension of the bracket including W, D, t, R and weld length *l*.

13.10.6 Pipe support bracket

The diagram, Figure 13.20, shows the layout of a hot-water return pipe for a high-rise office block. The constraints for this design problem arise out of the need to support this pipe entirely on the two brackets shown, with approximately zero load at the pump flange.

Design the brackets to carry the applied load.

Note
Use steel with S_y = 210 MPa; sheer strength S_s = 52.5 MPa; bearing strength = 105 MPa (average contact pressure). Refer to Figure 13.20.

The expansion joint provides no restraint for the pipe system. In fact assume that there are no external bending loads applied on the pipe system.

Neglect dynamic loads, but comment on what these may be due to and how you would assess their magnitude in a practical situation.

The pipe is 150 mm outside diameter steel.

13.10.7 Roof truss support bracket

Figure 13.21 shows a detail of the roof structure for a building. Your attention is directed at the mild steel brackets used to support the end of each rafter and make the connection between the rafter and the roof truss. This bracket, called a 'boot', is to be constructed from 6 mm sheet steel bent to shape and welded. You are responsible for the specification of the weld and other joint details concerned with connecting the support boot to the roof truss. In your design specification you must:

1. enumerate the modes of failure the boot should be designed to resist;
2. specify weld details;
3. specify bolts required for connecting boot to roof truss;
4. produce a dimensioned sketch of the boot, suitable for production.

Note
Use material specified as AS 1442 - CS1020 commercial grade mild steel.

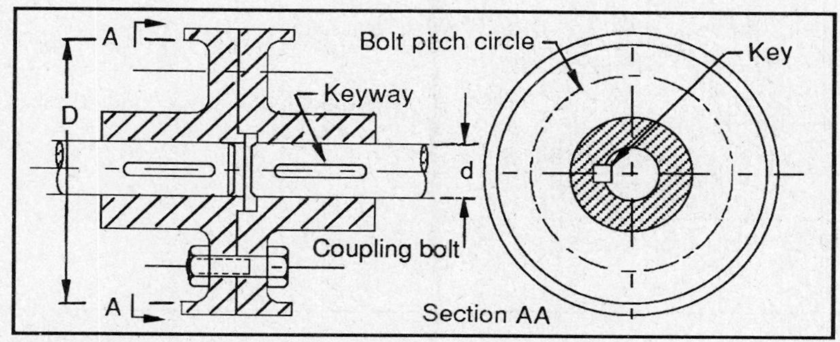

Figure 13.18 Flanged coupling

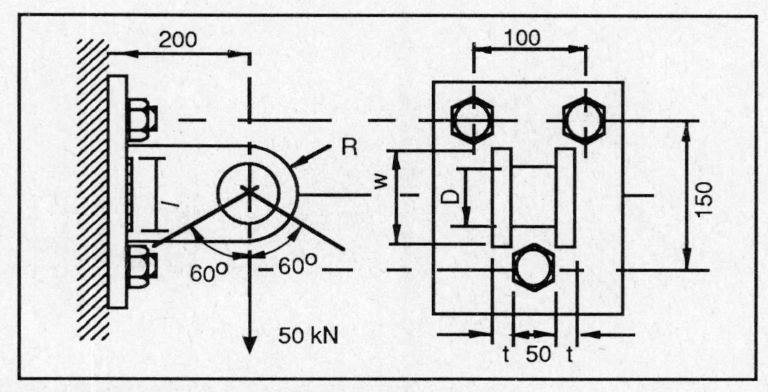

Figure 13.19 Welded bracket. All dimensions in mm

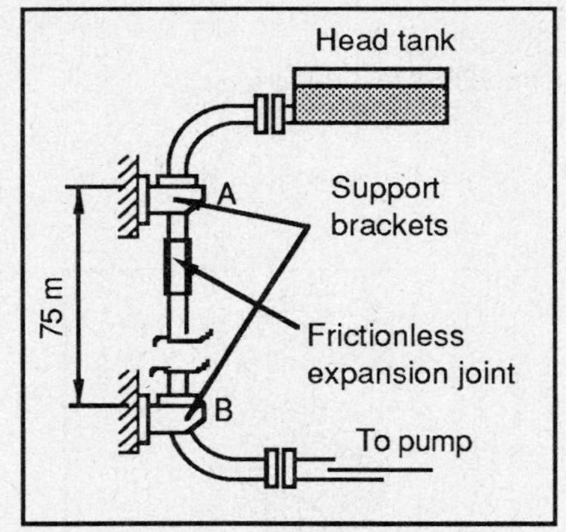

Figure 13.20 Pipe support brackets

The expansion joint provides no restraint for the pipe system. In fact assume that there are no external bending loads applied on the pipe system.

Neglect dynamic loads, but comment on what these may be due to and how you would assess their magnitude in a practical situation.

The pipe is 150 mm outside diameter steel.

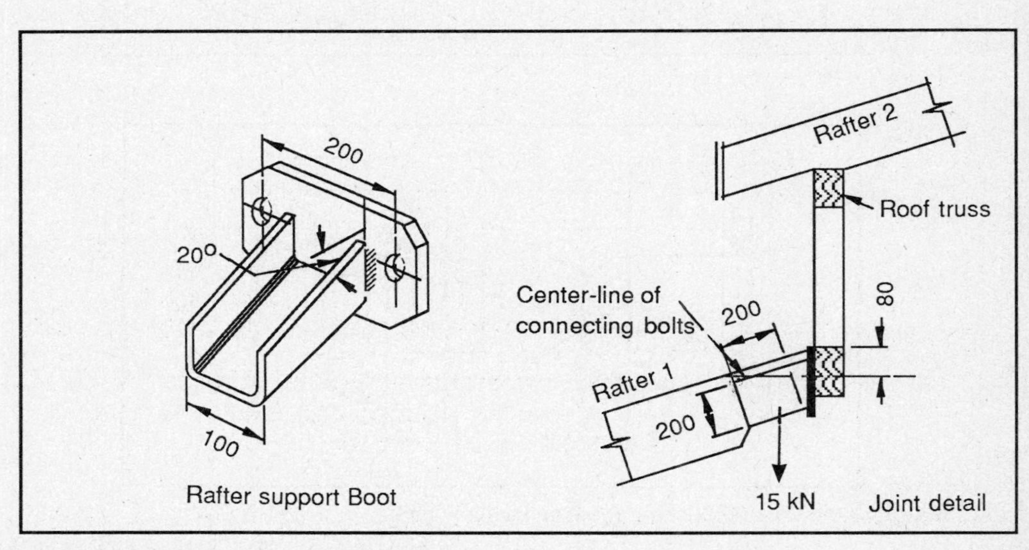

Figure 13.21 Roof truss support bracket

Chapter 14

Contactual elements

Very few things happen at the right time, and the rest do not happen at all. The conscientious historian will correct these defects.

Herodotus

Concepts introduced	elements subject to triaxial stress conditions; Hertzian stress fields.
Methods presented	evaluation of stress fields for solid elastic spheres and cylinders in contact.
Application	design of involute gears.

14.1 Introduction

Solid elastic bodies in contact occur in many engineering applications such as anti-friction or rolling element bearings, gear teeth and roller supported machine guideways. The state of stress in these bodies is triaxial. An elementary treatment of these stresses is given in Shigley, *Mechanical Engineering Design*. A more complete treatment is available from *Handbook of Engineering Mechanics* (ed. W. Flugge, McGraw Hill, 1962) and from the New Departure Division of General Motors Corporation, *Analysis of Stresses and Deflections for Solid Elastic Bodies in Contact*. Any serious study of the subject should include reference to this latter text.

As well as this remarkable text, students of this subject are referred to the classic book of Arvid Palmgren, *Ball and Roller Bearing Engineering* (1959), which was written while Dr. Palmgren worked for the Swedish A. B. Svenska Kullagerfabriken (SKF) company.

The topic of rolling element bearing design is in itself a fascinating field of conjecture about the apparent behaviour of the elements in elastic contact. Energy at the contact point is lost through imperfect restitution of the elastic energy of deformation (hence coefficient of restitution < 1) as well as due to the rolling element having to climb the 'hill' or 'bow wave' produced by the contact. Figure 14.1, reproduced from Palmgren's text, shows this effect. Further loss of energy results from the irreversible shearing of the lubricating fluids, as well as a 'gyratory' or induced rotary motion of the balls in ball bearings. All of these energy losses are usually handled by a single friction coefficient term resulting in the bearing having an efficiency of something less than 100%. These losses usually amount to no more than 2 or 3 percent.

The two special cases of contact to be considered in this introductory text are:

1. stresses and deformations for two spherical bodies in contact;
2. stresses and deformations for two cylindrical bodies in contact.

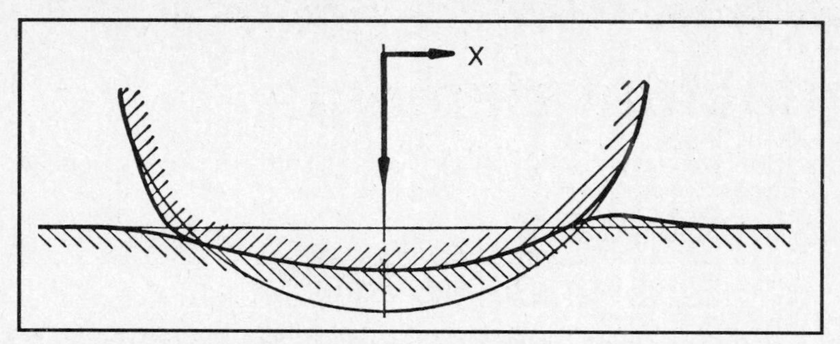

Figure 14.1 Ball rolling on flat surface, elastic deformation only
Source: Palmgren (1959).

14.2 Spherical bodies in contact

When two solid, elastic, curved bodies are pressed together under load, some flattening occurs in the neighbourhood of the contact point. The shape of the flat area of contact depends on the principal radii of curvature of the two contacting bodies. Hertz (1896) determined the distribution of pressures in the contact area and most texts refer to these (elastic) stresses as 'Hertzian stresses'. Later investigators have explored the whole near- field of stress regions in bodies contacting under load, but in this text we restrict our discussions to Hertzian stresses only.

Consider two contacting ellipsoidal bodies, body a and body b, with principal radii of curvature at the contact point of $R_{a1}, R_{a2}, R_{b1}, R_{b2}$, under the action of a normal load F. The resulting contact surface will be elliptical in projection, with semi-axes a and b. Over this area the pressure distribution will be ellipsoidal, with the maximum at the center of the ellipse. Hertz calculated the stresses and the semi-axes in terms of the elastic constants of the materials of the bodies in contact (elastic modulus and Poisson's ratio) and in terms of transcendental functions related to the elliptic integrals (refer to Jahnke and Emde, *Tables of Functions*, 1943).

For spherical bodies in contact, we can write:

$$R_{a1} = R_{a2} = a$$
$$R_{b1} = R_{b2} = b$$

The contact surface is circular in shape when projected into the plane and has a diameter $D_c = (2R_c)$ given by:

$$R_c^3 = \frac{3F}{8}\left[\frac{1-\mu_1^2}{E_1} + \frac{1-\mu_2^2}{E_2}\right]\left(\frac{a}{2} + \frac{b}{2}\right)^{-1}$$

where E is Young's modulus and μ is Poisson's ratio. The pressure distribution over the contact area is semi-elliptical in both solids. The maximum pressure occurs at the center of the contact area and is given by:

$$p_{max} = -\frac{3F}{2\pi R_c^2}$$

Notably these equations apply for a ball resting on a plane surface (i.e. as the radius a or b tends to ∞). Shigley (1986) gives charts of stress distributions as a function of distance from the contacting surface for steel with a Poisson's ratio of 0.3. The maximum shear stress in the material for this special case occurs at approximately $0.3R_c$ below the surface and is equal to approximately $0.3p_{max}$.

14.3 Cylindrical bodies in contact

The case of cylinders in contact is treated differently from the case of spheres. The stress field is no longer triaxial, since it does not vary parallel with the cylinder axis. Nonetheless the concepts are similar and the contact surface is rectangular in plane projection, having a width $2b$ as shown on Figure 14.2.

To illustrate the procedure for designing contactual elements, consider the case of two cylinders of length l in contact under the action of a transverse force F. So long as the deformation is elastic, the area of contact is a narrow rectangle of length l and half width b. This half width b is found from the theory of mechanics of solids to be:

$$b = \sqrt{\left\{\frac{2F}{\pi l}\left(\frac{1-\mu_1^2}{E_1} + \frac{1-\mu_2^2}{E_2}\right)\left(\frac{1}{d_1} + \frac{1}{d_2}\right)^{-1}\right\}}$$

where the suffixes 1 and 2 refer to the two cylinders and other symbols have their usual significance. For the usual engineering materials of construction, $\mu_1 = \mu_2 = 0.3$. The contact pressure has an elliptical distribution over the width $2b$, the maximum pressure being:

$$p_{max} = -\frac{2F}{\pi b l}$$

The shear stress reaches a maximum value just below the surface (approximately distance b below the surface), where τ_{max} is closely approximated by $0.3\ p_{max}$. The design of contactual elements under steady load is based on keeping p_{max} or τ_{max}

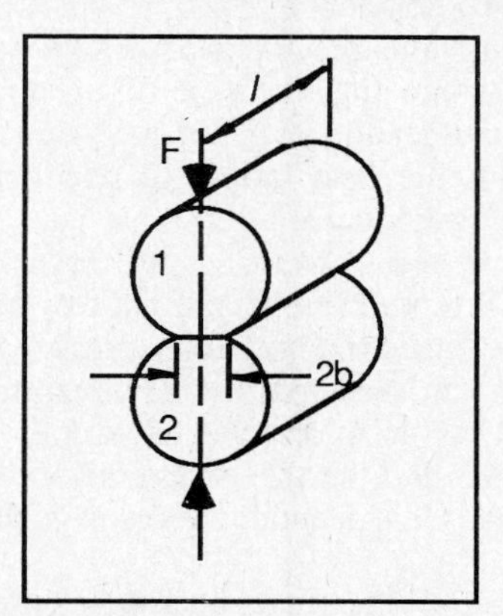

Figure 14.2 Cylindrical elements in contact

within allowable limits. The above results can be applied to a cylinder in contact with a plane surface by putting $d_2 = \infty$.

14.4 Cyclic loading

When two cylindrical surfaces roll or roll and slide against one another with sufficient force, a pitting failure will occur after a certain number of cycles of operation. The ability to resist this type of failure depends on the contact stresses, the number of cycles of operation, the surface finish and hardness of the materials used, the degree of lubrication, and the temperature of operation. While the mechanism of pitting is complex, some authorities postulate that a surface fatigue failure is initiated by a crack which orginates from the zone of maximum shear stress and then propagates to the surface. Eventually a small chip of material breaks away from the surface.

For design another material property is introduced, the surface endurance limit S_{fe}. This is the maximum contact pressure $\left(\frac{2F}{\pi b l}\right)$ which after 10^8 cycles of operation will cause failure by surface damage. Such failures are often referred to as wear, but should be distinguished from abrasive wear. Experimental results have also been reported in terms of a 'load-stress factor' K where:

$$K = \frac{F}{l}\left(\frac{1}{r_1} + \frac{1}{r_2}\right)$$

and K is a function of S_{fe} (Shigley, 1986). Given either K or S_{fe}, the cylinder dimensions — radii and length — can be chosen to ensure a satisfactory fatigue life.

14.5 Application to gear design

Gear design is governed by the two modes of failure to which gear teeth are subject:

1. tooth breakage, fracture due to excessive bending load; and
2. tooth wear, surface pitting due to excessive contact pressure.

In (1) the tooth is considered as a short cantilever and in (2) a pair of mating teeth as two bodies in rolling and sliding contact.

The British and American standards for gear design have been developed on this basis. The designer has to ensure that the size of the tooth, its face width and the materials used are such that neither the gear nor the pinion fail in bending or in wear. Usually wear of the pinion is the first failure to occur and this mode of failure is therefore given priority in the design calculations.

Design equations relating tooth bending and wear to the torque transmitted incorporate factors based on the proportions and radii of curvature of the profile used. Empirical factors are also introduced to allow for vibration and shock loads arising from inaccuracies in manufacture and assembly. Gear design standards and handbooks tabulate these factors, as well as the properties of various materials of construction in terms of 'bending stress factors' and 'surface stress factors'. Publication No. 68040A of the Engineering Sciences Data Unit, London, gives a good overview of the design of spur and helical gears.

The theory and practice of rolling contact bearings is presented in Houghton (1976). See Neale (1973) for useful information on contact phenomena and contact stresses.

Tables of data

1 Bending moments and deflections of beams

1 Cantilever - point load

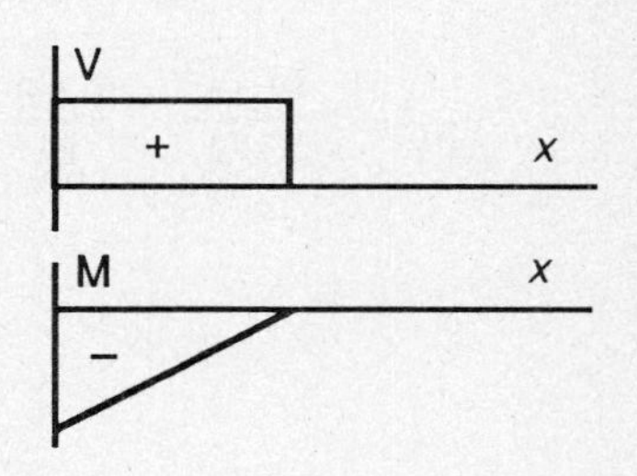

$R = V = F$ $\qquad$ $M = -\,Fa$

$M_{AB} = F(x - a)$ $\qquad$ $M_{BC} = 0$

$y_{AB} = \dfrac{Fx^2}{6EI}(x - 3a)$ $\qquad$ $y_{BC} = \dfrac{Fa^2}{6EI}(a - 3x)$

$y_{max} = \dfrac{Fa^2}{6EI}(a - 3l)$

2 Cantilever - distributed loadload

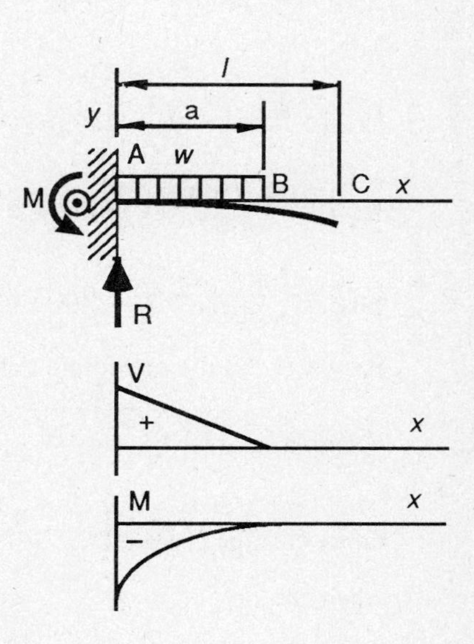

$R = wa$ $\qquad$ $M_A = -\,\dfrac{wa^2}{2}$

$V = w(a - x)$ $\qquad$ $M_x = -\,\dfrac{w}{2}(a - x)^2$

$y_{AB} = \dfrac{wx2}{24EI}(4ax - x^2 - 6a^2)$

$y_{BC} = y_{AB} + (l - a)\tan\theta_B$

$\tan\theta_B = \left(\dfrac{dy}{dx}\right)_B = -\,\dfrac{wa^3}{6EI}$

$y_{BC} = -\,\dfrac{wa^4}{8EI} - \dfrac{wa^3}{6EI}(l - a)$

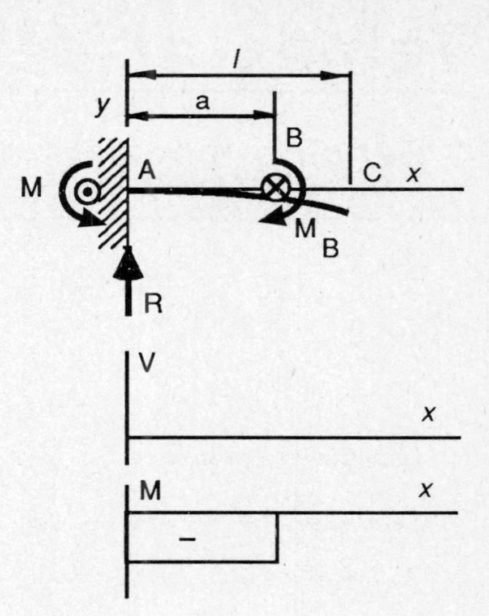

3 Cantilever - applied moment

$R = 0 \qquad M = M_B \qquad V = 0$

$$y_{AB} = -\frac{M_B x^2}{2EI}$$

$$y_{BC} = y_{AB} + (l - a)\tan\theta_B$$

$$\tan\theta_B = \left(\frac{dy}{dx}\right)_B = -\frac{M_B a}{EI}$$

$$y_{BC} = -\frac{M_B a^2}{2EI} - \frac{M_B a}{EI}(l - a)$$

4 Simply supported beam - point load

$$R_1 = \frac{F(l - a)}{l} \qquad R_2 = \frac{Fa}{l}$$

$$V_{AB} = R_1 \qquad V_{BC} = R_2$$

$$M_{AB} = \frac{Fx}{l}(l - a) \qquad M_{BC} = \frac{Fa}{l}(l - x)$$

$$y_{AB} = \frac{Fx(l - a)}{6EIl}(x^2 - 2la + a^2)$$

$$y_{BC} = \frac{Fa(l - x)}{6EIl}(x^2 - 2lx + a^2)$$

for $a < (l - a)$ the maximum deflection

occurs at $x = l - \sqrt{\frac{l^2 - a^2}{3}}$

$$y_{max} = -\frac{Fl^3}{48EI}(16n)\left\{\frac{1 - n^2}{3}\right\}^{1.5},$$

where $a = nl$

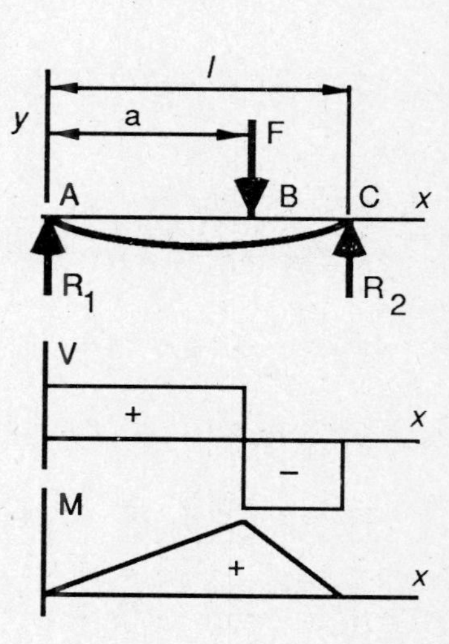

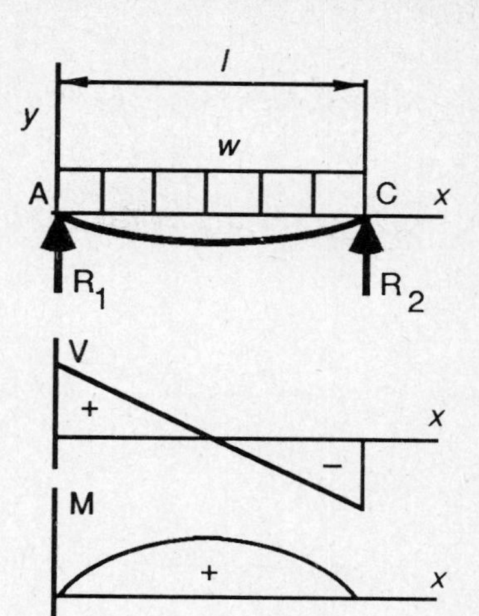

5 Simply supported beam - distributed load

$R_1 = R_2 = \frac{wl}{2}$ $\quad V = \frac{wl}{2} - wx$

$M = \frac{wx}{2}(l - x)$

$y = \frac{wx}{24EI}(2lx^2 - x^3 - l^3)$

$y_{max} = -\frac{5wl^4}{384EI}$

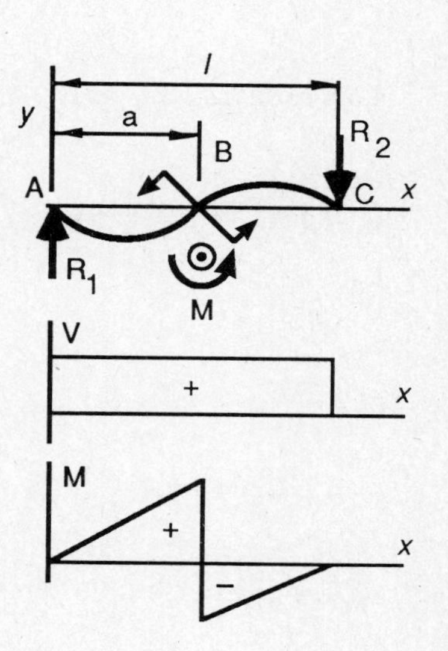

6 Simply supported beam - applied moment

$R_1 = -R_2 = \frac{M_B}{l}$ $\quad V = \frac{M_B}{l}$

$M_{AB} = \frac{M_B x}{l}$ $\quad M_{BC} = \frac{M_B}{l}(x - l)$

$y_{AB} = \frac{M_B x}{6EIl}(x^2 + 3a^2 - 6al + 2l^2)$

$y_{BC} = \frac{M_B}{6EIl}[x^3 - 3lx^2 + x(2l^2 + 3a^2) - 3a^2 l]$

7 Beam on fixed supports - point load

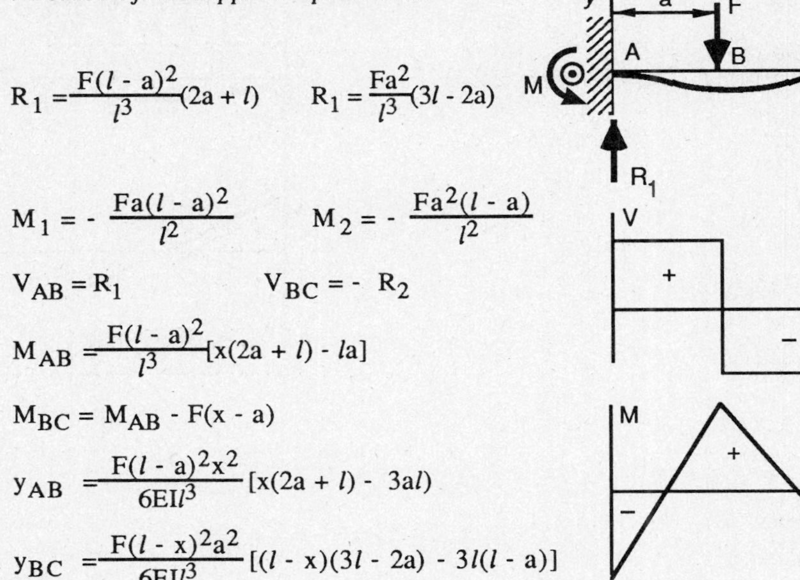

$$R_1 = \frac{F(l - a)^2}{l^3}(2a + l) \qquad R_1 = \frac{Fa^2}{l^3}(3l - 2a)$$

$$M_1 = -\frac{Fa(l - a)^2}{l^2} \qquad M_2 = -\frac{Fa^2(l - a)}{l^2}$$

$$V_{AB} = R_1 \qquad V_{BC} = -R_2$$

$$M_{AB} = \frac{F(l - a)^2}{l^3}[x(2a + l) - la]$$

$$M_{BC} = M_{AB} - F(x - a)$$

$$y_{AB} = \frac{F(l - a)^2 x^2}{6EIl^3}[x(2a + l) - 3al)$$

$$y_{BC} = \frac{F(l - x)^2 a^2}{6EIl^3}[(l - x)(3l - 2a) - 3l(l - a)]$$

8 Beam on fixed supports - distributed load

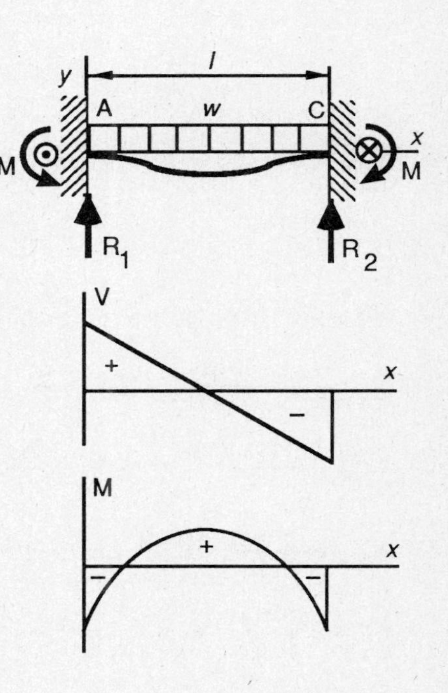

$$R_1 = R_2 = \frac{wl}{2} \qquad M_1 = M_2 = -\frac{wl^2}{12}$$

$$V = \frac{w}{2}(l - 2x)$$

$$M = \frac{w}{12}(6lx - 6x^2 - l^2)$$

$$y = -\frac{wx^2}{24EI}(l - x)^2$$

$$y_{max} = \frac{wl^4}{384EI}$$

2 Conversion tables

1 Measurement prefixes shown as powers of ten (10^m)

Prefix	Symbol	Multiplier
Tera	T	10^{12}
Giga	G	10^9
Mega	M	10^6
kilo	k	10^3
centi	(c)	10^{-2}
mili	m	10^{-3}
micro	μ	10^{-6}
nano	n	10^{-9}
pico	p	10^{-12}
femto	f	10^{-15}
atto	a	10^{-18}

2 Conversion factors for some common imperial units to SI units

Quantity (units)	common symbol	from imperial	to SI	multiplier
Length(L)	l	inch(in)	metres(m)	25.4×10^{-3}
		feet(ft)	m	0.3048
		yard	m	0.9144
		mile	km	1.609×10^3
Area(L^2)	A	in^2	m^2	6.451×10^{-4}
		ft^2	m^2	9.29×10^{-2}
Volume(L^2)	V	in^3	m^3	16.39×10^{-6}
		ft^3	m^3	28.32×10^{-3}
		gallon(U.S.)	m^3	3.785×10^{-3}
Mass(M)	m	ounces (avoirdupois)	kg	28.35×10^{-3}
		pounds(lb)	kg	0.453 592 37
Dynamic moment inertia(ML^2)	I,J	lb in^2	kg m^2	292.6 x 10-6
Density(ML^{-3})	ρ	lb/in^3	kg m^{-3}	27.68×10^3
		lb/ft^3	kg m^{-3}	16.018 5
		slug/ft^3	kg m^{-3}	515.379
Force(MLT^{-2})	F	lbf	Newton	4.448 222
Pressure	P	lb/ft^3	N m^{-2} (Pascal)	47.880 3
Stress	σ	lb/in^2(p.s.i.)	k Pa	6.894 757
Power	P	horsepower (hp)	kilowatt (kW)	0.745 699

Acceleration due to gravity is 9.806 65 m s^{-2}

Design workbook

Problems and questions to test your understanding of the science and art of engineering design.

1. *Theory of design*

1.1 A statement of the objectives of a university course in engineering would cover the following matters:

(i) insight into the nature of existing problems society expects professional engineers to solve;
(ii) vision to forecast the nature of future problems;
(iii) knowledge of relevant theory and practice;
(iv) competence in the application of that knowledge;
(v) communication of solutions to existing problems;
(vi) persuasion to gain acceptance of new solutions to new problems.

(a) What technical problems associated with your branch of engineering do you expect society to ask you to solve in the next 15 years?
(b) From what sources will you obtain ideas for solving these problems:
   (i) scientific theory;
   (ii) experimental research;
   (iii) personal ingenuity;
   (iv) successful engineering precedent;
   (v) some combination of these and possibly other sources.

   Give reasons for your answer.

1.2 Construct a chart showing the flows of information between the major activities in which a professional engineer may be engaged. Show on this chart how you would classify the work of engineers engaged in the following activities:

(a) designing an early warning device for level crossings;
(b) developing new alloy steels to resist hydrogen embrittlement in equipment for producing oil from coal;
(c) correcting errors in a computer program being installed to sequence the operation of traffic lights and thus assist traffic flows in a busy part of a large city;
(d) developing a mathematical model for the airshed in the region around a large coal-fired power-station to enable predictions to be made of the rate of dispersal of atmospheric pollutants;
(e) building a hi-fi set.

Briefly give the reasons for your classifications.

1.3 A community need or concern, perhaps vaguely apprehended in the first instance, leads via processes of information search and idea generation to the identification and specification of technical problems, and hence to proposals for their solution. Thus the stimulus to engineering problem solving is provided by statements such as the following:

'The demand for electric power where we live is doubling every 8 to 10 years.'
'My clothes are dirty.'
'The time the average person spends travelling to and from work is excessive.'
'Too many people are killed and injured when travelling on our roads.'
'Vandals are causing a lot of damage to telephone booths.'
'Research has shown that the efficiency of clerical workers falls off rapidly as the ambient temperature increases above 27 degrees Celsius.'

Identify technical problems to which the above statements give rise, and show how the problems are derived from the statements. Investigate the characteristics of the technical problems you have identified and the extent to which they relate to the social and economic well-being of society and its potential for future growth.

1.4 A 'design tree' is a convenient way of representing in visual form the structure of a complex design problem and the inter-relationships between the sub-problems of which it is composed and also the sub-solutions to which they give rise.

(a) Draw design trees for:
  (i) insulators for electric fences;
  (ii) new machinery for harvesting apples;
  (iii) a complex engineering system or device of your own choice.
(b) Show by means of a design tree or similar hierarchical diagram how research into road safety could (or should) be organized.
(c) Compare the diagrams drawn in response to (a) and (b) and highlight any differences between them.

1.5 The bridge is an important part of a ship's control system.

(a) Draw a block diagram, similar to that in Figure 1.6, to show the operation of a ship as an engineering system.
(b) Identify and list the types and items of information that flow into and out of the bridge on a ferry which transports people and vehicles across a 50 km stretch of water.
(c) Comment on the influence of these information flows on the design and layout of the bridge on a ship of this type.

2. *Problem definition*
2.1
(a) Suppose that you are responsible for the design of a new vehicle for a local transport authority. Which of the items listed below would you classify as:

  (i) a design objective you are trying to satisfy;
  (ii) a design criterion you would use to compare the merits of your new design with those of existing vehicles; or
  (iii) something else?

  If you classify an item under (iii), specify what that 'something else' is:

  (1) comfort of seated passengers;
  (2) height of steps at entrance to vehicle;

(3) number of people capable of being transported by the vehicle;
(4) ease of collecting fares;
(5) maximum horizontal acceleration ;
(6) ease of maintenance;
(7) width of passenger seats;
(8) number of breakdowns expected.

(b) In order to establish the 'best' choice from available alternatives the designer uses measures called 'performance criteria'. Clearly these criteria must be identified before the choice is made. List performance criteria for the following:
(i) a new washing machine to compete on international markets;
(ii) a new freeway linking the inner city to outer suburbs;
(iii) some entertaining event in the bicentenary celebrations of a nation.

2.2 Engineering design may embrace very complex situations. In order that you may deal with such situations in an effective way you have been shown a set of artifices or techniques for reducing the level of difficulty encountered so that the problems are more readily managed. Consider the following problem statements and imagine yourself as the engineer who has to devise effective responses to them:

'make our public transport efficient';
'design an improved emission control device for automobiles';
'find an appropriate site for a new power station near a large city';
'redesign an old lecture theatre for contemporary use';

For each of the four problem statements:

(a) state (with reasons) whether the statement is problem-oriented or solution-oriented;
(b) generate a solution strategy and show by means of a flow chart how it would be implemented;
(c) indicate which weapons in your intellectual armory would be brought into action by your strategy.

Note: Do not solve the problems, show how you might best reach acceptable solutions.

2.3 The preparation of an initial appreciation of a new design problem was described in Section 2.2.6 of Chapter 2 and an example was given. The format is as follows:

(i) objectives to be satisfied and their relative importance;
(ii) criteria for evaluating alternative proposals;
(iii) information required as a basis for the design as input to the processes of idea generation and decision making; possible sources from which this information might be obtained;
(iv) problems foreseen to arise in the course of the design;
(v) plan of campaign for solving these problems in an orderly manner.

For each of the cases set out below prepare an initial appreciation of the task you have been asked to perform.

(a) You are an engineer with a local transit authority and have been asked to design a new system for automatically cleaning the external surfaces of its vehicles.
(b) You are a consultant hired to design the casualty section of a large new public hospital to be built near a heavily populated city area. Your investigations and

recommendations are to include the layout of emergency admission and treatment areas as well as hospital procedures and management systems.

(c) You are an engineer in a city administration and have been asked to design a system for controlling traffic lights to give optimum traffic flow in the Central Business District.

(d) You are an engineer employed by your university and have been asked to design a cabinet for displaying a set of important and irreplaceable documents donated to the university. The documents are of great historical interest and consist of hand-written letters by leading citizens concerning their plans for the founding of the university. The letters cover 20 pages of various sizes, with writing on both sides of most pages.

3. *Generation of ideas*

3.1 The fuel consumption of a motor car is proportional to its mass. List the vehicle design changes you would propose to a car manufacturer that would lighten a vehicle yet retain necessary structural integrity. Illustrate your ideas with sketches.

3.2 Elucidate the structure of design problems associated with the following matters and apply the methods of systematic combinatorial search to find as many possible solutions as you can:

(a) campus notice boards;
(b) tennis racquet;
(c) sandwich;
(d) jogging shoes.

3.3 Survey engineering journals and magazines to find examples of the application of design methods described in Chapter 2. Write concise case histories of two examples you find the most interesting.

3.4 Investigate to find as many instances as you can of structures and devices in nature which are analogous to engineering structures and machines.
Hint: A useful starting point for this exercise is found in Professor Michael French's book *Invention and Evolution — Design in nature and engineering*, Cambridge, 1988.

3.5 Think up new ways of teaching and learning in a university subject. Try out your ideas on:

(a) two of your friends: and
(b) a university teacher.

Analyze their responses for evidence of 'rule-outs'. Write a short report setting out your proposals and your analysis of the responses you received.

3.6 Dripping paint brushes and paint tins present problems in practice. Prepare a log recording all your ideas for defining and solving these problems. Try to record how much information you can generate in a given time, say ten minutes.

3.7 Generate ideas for solving the design problems in Question 2.3 under 'Problem Definition', that is, those problems for which initial appreciations have been prepared.

4. *Decision making*

4.1 In your answer to an earlier question you proposed ways of lightening automobiles. Using the format set out in Section 3.3 of Chapter 3 construct a decision table and show how it would be used to select the most promising proposal(s).

4.2 Several alternative sources of power are available for telecommunication repeater stations in remote country districts — solar cells, wind turbines, diesel generator sets. Draw up a decision table to show how you would decide which source to use in a particular application. As far as you are able, enter appropriate information into the cells in the table and arrive at a specific recommendation.

4.3 Consider the problem of designing passenger carriages, say for a suburban railroad system. Given that the overall size of the carriage has been determined, an important design decision then concerns the allocation of floor space. What proportion of floor space should be devoted to seating and what proportion to standing room in aisleways and exits? How would you recommend that the designer make this decision? In other words, if you were the responsible person how would you decide to allocate floor space? Give a detailed discussion of the decision making process. Identify relevant design variables and the scales on which they would be measured.

4.4 Suppose you are a senior engineer with an electricity generating company and are charged with the responsibility of recommending to the board of management the size and timing of construction of new power stations. List the factors upon which your recommendations would be based. Identify all relevant sources of uncertainty, and explain how you would cater for these uncertainties when preparing your recommendations.

4.5 Suppose you are a planning engineer with your local telephone company and have been asked to investigate the future impact of optical fibers on operations over the next ten years. Identify the uncertainties which would affect this planning exercise and the predictions stemming from it. Describe how you would handle these uncertainties in compiling and presenting a report for management.

5. *Economic evaluation*

5.1 You are a consultant to a large corporation whose annual energy bill is more than $2 million. You estimate the cost of a thorough energy survey to be about $60,000. With a prime lending rate of 17 percent, what annual savings should you be able to achieve if the consultant's survey fee is to be recovered in less than six years?

5.2 Your employer is considering the installation of a new widget machine. The cost of installation is expected to be $50,000 with an estimated operating life of ten years. A ten-year old machine costs approximately $10,000. Widget sales estimate is $10,000 in the first year with inflation-indexed income increasing in each subsequent year at a rate of 15% per annum. Current bank lending rate for investment is 17% p.a. Estimate the simple payback period.

5.3 Expansions to a petrochemical plant require the purchase of a special 200 kW electric motor to drive a new gas compressor, the rated motor output being 200 kW. Three motor manufacturers have submitted tenders as follows:

Manufacturer	A	B	C
Cost of motor delivered to site	$ 28,875	$ 34,870	$ 36,520
Guaranteed motor efficiency	0.91	0.92	0.93

The plant will be operated for 100 hours a week, 50 weeks a year, for ten years. At the start of the 10 year period it is predicted that electrical power will be available at a cost of 3.0 cents per kWh, increasing by equal annual increments to 4.8 cents per kWh at the start of the tenth year. Electricity bills are paid quarterly.

(a) On the basis of the information given which motor would you recommend be purchased?
(b) Is your answer to (a) sensitive to variations in the cost of power, for example, if power costs were to double over the ten year period to 6 cents per kWh, would this affect your recommendation?
(c) What additional information would you seek in order to reach a final decision in this matter?

5.4 An extract is given below from a consultant's report in which an economic evaluation is made of a proposal to reduce energy and maintenance costs in a university building by replacing existing pipelines.

(a) Derive the following equation which is quoted without proof in the extract:

$$PW = AS\frac{1+p}{r-p}\left\{1-\left(\frac{1+p}{1+r}\right)^n\right\}$$

where: PW = present worth equivalent to annual savings of AS dollars;
n = number of years;
p = annual rate of increase of energy and maintenance costs;
r = interest rate.

(b) Show that the simple payback period for a capital investment of \$85,000 is 9.4 years, as stated in the extract, for the case where $p = 0.1, r = 0.1$.

Extract:

The present worth of these savings over 15 years provides an indication of whether replacement of existing lines is an economic proposition compared with investing the capital amount at current interest rate of ten percent.

The present worth of the anticipated savings is given by the formula (assuming that energy and maintenance costs rise by the same percentage each year):

$$PW = AS\frac{1+p}{r-p}\left\{1-\left(\frac{1+p}{1+r}\right)^n\right\}$$

where PW: = present worth equivalent to annual savings of AS dollars;
n = number of years;
p = annual rate of increase of energy and maintenance costs =10 percent;
r = interest rate.

Thus the present worth of the \$9,000 p.a. savings is \$135,000 compared with the capital cost of \$85,000 to replace pipelines, which suggests that replacement of the piping would be an economic proposition. The following table sets out the present worth of the anticipated savings for different values of p and n and confirms that replacement of the pipelines would be an economic proposition for lower rates of increases in energy and maintenance costs over a 15 year period (see shaded area in table). Values are in whole dollars.

p %	n years			
	5	10	15	20
0	34,100	55,300	68,500	76,600
4	38,200	67,000	88,700	105,210
6	40,320	73,800	101,700	124,800
10	45,000	90,000	135,000	180,000

6. *Benefit — Cost analysis*

6.1 Assess the benefits to be derived from and the costs associated with legislation restricting the emission of pollutants from the exhausts of automobiles. Your assessment should include quantitative evaluations of the consequences of implementing such legislation, and it should be in sufficient depth for you to make a judgment on just what legislation you would support as being in the general interest of society.

6.2 Assess the benefits to be derived from and costs associated with legislation for the compulsory wearing of seat-belts by (a) drivers and (b) both drivers and passengers of automobiles and trucks. Your assessment should include quantitative evaluations of the consequences of implementing such legislation, and it should be in sufficient depth for you to make a judgment on just what legislation you would support as being in the general interest of society.

7. *Ergonomics*

7.1 Prepare a design audit, that is, a critical review with constructive comments, of the lecture theatres you have used in your university (plenty of sketches please).

7.2 Computer rooms typically contain terminals, printers, and associated equipment. Constructively criticize the layout of a computer room with which you are familiar, also the design of the equipment in it from the point of view of ergonomics, that is, paying attention to the needs and characteristics of the people using the room (plenty of sketches, please).

7.3 Consider the work of drivers of diesel locomotives on long-distance (for example, inter-city and inter-capital) trains. What characteristics of their work justify describing it as a 'vigilance' task ? What steps have been/should be taken to ensure that driver alertness is maintained ?

7.4

(a) Describe how people respond to informational inputs in situations where they are operating and controlling complex systems and where for varying periods of time they receive:

(i) little information on which they have to act, for example, the operator in control rooms of power stations ;

(ii) lots of information, for example, the drivers of automobiles at busy intersections.

(b) In the light of your answers to part (a) and from your knowledge of ergonomics, state the advice you would give to the designer of:

(i) a control room for a power station ;
(ii) busy traffic intersection.

7.5 Optimize the design of the driver's compartment in a family, 4-door, 6-cylinder, automobile from the point of view of ergonomics to suit a wide range of drivers of both sexes. Consider driver position and comfort, also the position, layout, and accessibility of displays and controls. Illustrate your ideas with sketches.

8. *Safety and failure analysis*
Note: An interesting article on engineering safety was published in the March 1980 issue of *Scientific American*, pages 33 to 45. This article is highly recommended; it contains examples of fault trees and event trees for nuclear reactors.

8.1
(a) What events or combinations of events could lead to a group of people being trapped in an elevator in, say, a university engineering building?
(b) By assigning probabilities to individual events and making appropriate estimates, predict the probability of an engineering student being trapped in an elevator at some time during his or her course.
(c) What actions could or should be taken to protect travellers in elevators from system (and other) failures?
(d) Show the logic of these actions on a failure analysis diagram.

8.2 Computers and data-processing equipment are often installed in laboratories in universities.

(a) What serious consequences would follow from an interruption to the power supply to one of these laboratories — an interruption for an extended period of time, say several hours ?
(b) What event or combination of events could lead to such a power failure?
(c) What action is taken or should be taken to protect a computer from a failure in power supply?
(d) Show the logic of this protective action on a failure analysis diagram.

8.3 Construct a fault tree or failure analysis diagram showing the inter-related sequences of events which would lead a student in the second or third year of an engineering course to fail in the examinations at the end of the year. Develop an equation which relates the probability of passing a year to the probabilities of a number of subsidiary events upon which final success depends.

8.4 Two extracts are quoted below from a local newspaper. Discuss these extracts in the context of modern accident theory. Compare and contrast the viewpoints adopted, particularly with respect to causes of road accidents.

Extract (1): Published under the headlines, 'Beware of Killers on Roads — Coroner's Easter Warning'.

'Irresponsible and impatient drivers are always killers on the roads', the City Coroner, Mr Pascoe, said today. Mr Pascoe knows about death and accidents. He has been coroner for 14 years and held inquests on hundreds of people. In that time he has only once found 'sheer accident' in a death in a road crash. 'In all other cases there has been negligence by some party involved', he said.

Extract (2): Report of an analysis of a fatal accident prepared by a government Traffic Accident Research Unit.

The crash happened in a semi-rural residential area on the outskirts of a major city. Two vehicles were involved. In one car, a station wagon, was a 60-year old man with 14 years' driving experience. He was heading for work, driving along quietly, with his seat belt on. 1.5 km down the road a 20-year old student was getting into his small car to take his aunt and cousin to the airport.

His middle-aged aunt got into the front seat and the cousin into the back. In spite of the little extra effort required to clip the belt up, the woman drew it across herself and held the buckle in her hand. But the driver put his belt on. There was no belt in the back seat so the cousin sat diagonally across the seat looking towards the driver.

They drove down the drive, and out into the street. The road was very narrow, pretty straight with some crests and hills. A 4-meter strip down the center was tarred and very few people drove quickly along it. Most people drove down the bitumen putting their left wheels into the gravel if another car came the other way. School children walked along the verge at that time of day so it was additionally prudent to drive slowly.

The trio were only 150 meters from their gate and heading up a slight hill in third gear when the station wagon came over the hill in the center of the road. The driver of the wagon could not pull off to his left as there were children on the verge. In fact the wagon driver had to move right. The car driver put on his brakes and swerved to the left. The two vehicles collided head-on.

The student's car was the more severely damaged. The cousin was hurled across the back seat, ending up half on the seat and half on the floor. The woman was thrown into the parcels shelf and then back on to the seat. The impact punctured her lungs and she died two minutes later. The student hit his head and was dazed. His only injuries were cuts to his eye, lips, and hands, and bruises across the chest and stomach where the seat belt had done its work. The wagon driver suffered bruised knees and belt bruises to his chest. A main factor in the chain that caused yet another road death was the narrow, hilly road barely wide enough for two cars to pass side-by-side.

The presence of the school children further reduced the effective width of the road. But the dominant link in the fatal chain was that the woman was holding the seat belt clip in her hand instead of buckling up.

References

Adams, J.L. *Conceptual Blockbusting.* Harmondsworth, Middlesex: Penguin, 1987.

Alexander, C. *Notes on the Synthesis of Form.* Cambridge, Mass.: Harvard University Press, 1964.

Alexander, C. and Manheim, M.L. "The Design of Highway Interchanges: An Example of a General Method for Analyzing Engineering Design Problems ", *Highway Research Record,* No. 83, 1964-65, pp. 48-87.

Allen, T.J. "Studies of the Problem Solving Process in Engineering Design", *I.E.E.E. Transactions on Engineering Management,* Vol. EM-13, 1966, pp. 72-83.

Archer, L.B. *Systematic Method for Designers.* London: Council of Industrial Design, 1965.

____" The Structure of the Design Process ", Chapter 8 in A. Ward and G. Broadbent (eds), *Design Methods in Architecture.* London: Lund, Humphries, 1969.

Ashford, F. *The Aesthetics of Engineering Design.* London: Business Books, 1969.

Asimow, M. *Introduction to Design.* Englewood Cliffs, N.J.: Prentice Hall, 1962.

Avallone, E.A. and Baumeister, T. (eds) *Marks' Standard Handbook for Mechanical Engineers* (Ninth edn). New York: McGraw-Hill, 1987.

Baumeister, T. (ed.) *Marks' Standard Handbook for Mechanical Engineers* (Eighth edn). New York: McGraw-Hill, 1978.

Black, H.C. "Safety, Reliability, and Airworthiness", Paper in A.M. Freudenthal (ed.), *Proceedings of International Conference on Structural Safety and Reliability.* Oxford: Pergamon, 1972.

Bloom, B.S. (ed.) *Taxonomy of Educational Objectives—Handbook I: Cognitive Domain.* New York: David McKay, 1956.

Borchardt, H. "Shortcuts for Designing Shafts", *Machine Design,* Vol. 45, No. 3, 1973, pp. 139-41.

Bridgwater, A.V. "Long Range Process Design and Morphological Analysis", *The Chemical Engineer,* Vol. 217, 1968, pp. CE 75-81.

British Standards Institution. *Publication PD 6112, Guide to the Preparation of Specifications.* London, 1967.

Cotgrove, S., Dunham, J. and Vamplew, C. *The Nylon Spinners.* London: George Allen & Unwin, 1971.

Crane, F.A.A. and Charles, J.A. *Selection and Use of Engineering Materials.* London: Butterworths, 1984.

de Neufville, R. and Stafford, J.H. *Systems Analysis for Engineers and Managers.* New York: McGraw-Hill, 1966.

de Bono, E. *Practical Thinking.* Harmondsworth, Middlesex: Pelican, 1976.

____ *Po: Beyond Yes and No.* Harmondsworth, Middlesex: Pelican, 1977.

De Garmo, E.P., Sullivan, W.G. and Canada, J.R. *Engineering Economy* (Seventh edn). New York: Macmillan, 1984.

Dixon, J.R. *Design Engineering: Inventiveness, Analysis, Decision Making*. New York: McGraw-Hill, 1966.

Dreistadt R. "An Analysis of the Use of Analogies and Metaphors in Science", *Journal of Psychology*, Vol. 68, 1968, pp. 97-116.

Dreyfuss, H. *The Measure of Man: Human Factors in Design*. New York: Whitley Library of Design, 1967.

Einstein, A. *Relativity—The Special and the General Theory*. London: Methuen, 1936, pp. 17-27.

Einstein, A. and Infeld, L. *The Evolution of Physics*. Cambridge, U.K.: Cambridge University Press, 1938, pp. 186-220.

Elmaghraby, S.E. "The Role of Modelling in Industrial Engineering Design", *Journal of Industrial Engineering*, Vol. 19, 1968, pp. 292-305.

Engineering Sciences Data Unit. *Publication 65004, Stress Concentration Data*. London, 1965.

____ *Publication 68040A, Design of Parallel Axis Spur and Helical Gears—hoice of Materials and Preliminary Estimate of Major Dimensions*. London, 1968.

____ Series On Stress and Strength—Vol. 5, *Data Sheets for Bolted Joints and Screw Threads*. London, 1965-86.

Faupel, J.H. and Fisher, F.E. *Engineering Design—A Synthesis of Stress Analysis and Materials Engineering* (Second edn). New York: Wiley-Inter science, 1981.

Field, B.W. *Some Computer-Aided Design Strategies*. Unpublished M.Eng.Sc. Thesis, University of Melbourne, 1970.

Flugge, W. (ed.) *Handbook of Engineering Mechanics*. New York: McGraw-Hill, 1962.

Fortini, E.T. *Dimensioning for Interchangeable Manufacture*. New York: Industrial Press, 1967.

French, M.J. *Conceptual Design for Engineers* (Second edn). London: The Design Council, 1985.

____ *Invention and Evolution—Design in Nature and Engineering:* Cambridge, U.K.: Cambridge University Press, 1988.

Freudenthal, A.M. (ed.) *Proceedings of International Conference on Structural Safety and Reliability*. Oxford: Pergamon, 1972.

Frichsmuth, D.S. and Allen, T.J. "A Model for the Description and Evaluation of Technical Problem Solving ", *I.E.E.E. Transactions on Engineering Management*, Vol. EM-16, 1969, pp. 58-64.

Frost, N.E., Marsh, K.J. and Pook, L.P. *Metal Fatigue*. Oxford: Oxford University Press, 1974.

Gagne, R.M. "Problem Solving and Thinking", *Annual Review of Psychology*, Vol. 10, 1959, pp. 147-72.

____ *The Conditions of Learning* (Fifth edn). New York: Holt, Rinehart and Winston, 1985.

Gasparini, G.S. and Chong, M.S. *Motor Car Side Pillars*. Design Report, Department of Mechanical Engineering, University of Melbourne, 1969.

George, F.H. *Models of Thinking*. London: George Allen & Unwin, 1970.

Gibbs-Smith, C.H. *Sir George Cayley's Aeronautics 1796-1855*. London: H.M.S.O., 1962.

Goldsmith, S. *Designing for the Disabled.* London: Royal Institute of British Architects, 1967.

Gordon, J.E. *Structures or Why Things Don't Fall Down.* Harmondsworth, Middlesex: Penguin, 1979.

____*The New Science of Strong Material or Why You Don't Fall Through the Floor.* Harmondsworth, Middlesex: Penguin, 1976.

Gordon, W.J.J. *Synectics—The Development of Creative Capacity.* New York: Harper and Row, 1961.

Gorenc, B. and Tinyou, R. *Steel Designers Handbook* (Fourth edn). Sydney: University of New South Wales, 1985.

Gott, H.H. and Berridge, D.R. "Designing a Large Power Station at Fawley", *Proceedings, Institution of Mechanical Engineers,* Vol. 180(3M), 1965-66, pp. 60-71.

Guilford, J.P. *The Nature of Human Intelligence.* New York: McGraw-Hill, 1967.

Gurney, T.R. *Fatigue of Welded Structures.* Cambridge, U.K.: Cambridge University Press, 1968.

Hadamard, J. *An Essay on the Psychology of Invention in the Mathematical Field.* Princeton, N.J.: Princeton University Press, 1949.

Hall, A.D. *A Methodology for Systems Engineering.* Princeton, N.J.: Van Nostrand, 1962.

Harary, F. *Graph Theory.* Reading, Mass.: Addison-Wesley, 1969.

Harvey, G.R. "The Optimum Design of a Tensile Stressing Attachment for Rectangular Section Concrete Specimens", *Bulletin of Mechanical Engineering Education,* Vol. 4, 1965, pp. 203-7.

Hertz, H. *Miscellaneous Papers.* English translation by D.E. Jones and G.A. Schott. New York: Macmillan, 1896.

Herzberg, F., Mausner, B. and Snydermann, B.B. *The Motivation to Work* (Second edn). New York: Wiley, 1959.

Heywood, J. "Qualities and Their Assessment in the Education of Technologists", *Bulletin of Mechanical Engineering Education,* Vol. 9, 1970, pp.,15-29.

Heywood, R.B. "Tensile Stresses in Loaded Projections", *Proceedings, Institution of Mechanical Engineers,* Vol. 159, 1948, pp. 384-98.

____ *Designing Against Fatigue.* London: Chapman and Hall, 1962.

Hill, P.H. *The Science of Design.* New York: Holt, Rinehart and Winston, 1970.

Hinton, B.L. "A Model for the Study of Creative Problem Solving", *Journal of Creative Behaviour,* Vol. 2, 1968, pp. 133-42.

Houghton, P.S. *Ball and Roller Bearings.* London: Applied Science Publishers, 1976.

Horger, O.J. (ed.) A.S.M.E. Handbook, *Metals Engineering - Design* (Second edn). New York: McGraw-Hill, 1965.

Jahnke, E. and Emde, F. *Tables of Functions with Formulae and Curves.* New York: Dover, 1943.

Johnson, R.C. *Mechanical Design Synthesis with Optimization Applications.* New York: Van Nostrand, Reinhold, 1971.

Jones, J.C. *Design Methods—Seeds of Human Futures.* London: Wiley-Interscience, 1970.

Kepner, C.H. and Tregoe, B.B. *The Rational Manager.* New York: McGraw-Hill, 1965.

Kratwohl, D.R., Bloom, B.S. and Masia, B.B. *Taxonomy of Educational Objectives—Handbook II: Affective Domain.* London: Longmans, 1964.

Krick, E.V. *An Introduction to Engineering and Engineering Design* (Second edn). New York: Wiley, 1969.

Langer, B.F. "Design of Vessels Involving Fatigue", Chapter 8 in R.W. Nichols (ed.), *Pressure Vessel Engineering Technology.* Amsterdam: Elsevier, 1971.

Lee, W., Christensen, J.H. and Rudd, D.F. "Design Variable Selection to Simplify Process Calculations", *American Institute of Chemical Engineers Journal,* Vol. 12, 1966, pp. 1104-10.

Lees, W.A. *Adhesives for Engineering Design.* London: Design Council, 1984.

Lewis, W.P. "Design in an Undergraduate Course in Mechanical Engineering", *Journal, Institution of Engineers, Australia,* Vol. 39, 1967, pp. 109-21.

____ "Subjective Assessments of Probability in the Design of Mechanical Systems", *Proceedings of International Conference on Engineering Design,* Hamburg, 1985, pp. 555-62.

Luckman, J. "An Approach to the Management of Design", *Operational Research Quarterly,* Vol. 18, 1967, pp. 345-58.

Mann, J.Y. *Fatigue of Materials.* Melbourne: Melbourne University Press, 1967.

Marples, D.L. "The Decisions of Engineering Design", *I.E.E.E. Transactions on Engineering Management,* Vol. EM-8, 1961, pp. 55-71.

Maslow, A.H. *Motivation and Personality* (Second edn). New York: Harper and Row, 1970.

McAree, P. "Planning a Range of Centrifugal Pumps Based on Cost Analysis and Using Computer Aid", *Proceedings, Institution of Mechanical Engineers,* Vol. 186, 1972, pp. 595-602.

McHarg, I.L. *Design with Nature.* Garden City, N.Y.: The Natural History Press, 1970.

McKee, R.B. *To Support a Mountain.* EDP Report No. 6-64, Department of Engineering, University of California, Los Angeles, California, 1964.

Megson, T.H.G. *Aircraft Structures for Engineering Students.* London: Edward Arnold, 1972.

Merrifield, P.R., Guilford, J.P., Christensen, P.R. and Fricke, J.W. "The Role of Intellectual Factors In Problem Solving", *Psychological Monographs,* Vol. 76, 1962, Whole No. 529.

Middendorf, W.H. *Engineering Design.* Boston: Allyn and Bacon, 1969.

Miller, G.A. "The Magical Number Seven, plus or minus Two: Some Limits on our Capacity for Processing Information", *Psychological Review,* Vol. 63, 1956, pp. 81-97.

Mitroff, I.I. "Simulating Engineering Design—a Case Study on the Interface between the Technology and Social Psychology of Design", *I.E.E.E. Transactions on Engineering Management,* Vol. EM-15, 1968, pp. 178-87.

Morgan, C.T., Cook, J.S., Chapanis, A. and Lund, M.W. *Human Engineering Guide to Equipment Design.* New York: McGraw-Hill, 1963.

Morgan, J. *A I D A—A Technique for the Management of Design.* London: Institute for Operational Research, 1971.

Murrell, K.F.H. *Ergonomics.* London: Chapman and Hall, 1965.

Nadler, G. "An Investigation of Design Methodology", *Management Science*, Vol. 13, Series B, 1967, pp. 642-55.

Neale, M.J. (ed.) *Tribology Handbook*. London: Butterworths, 1973.

New Departure Division of General Motors Corporation. *Analysis of Stresses and Deflections for Solid Elastic Bodies in Contact*. Bristol, Connecticut: New Departure, 1946.

Nichols, R.W. (ed.) *Pressure Vessel Engineering Technology*. Amsterdam: Elsevier, 1971.

Niebel, B.W. and Draper, A.B. *Product Design and Process Engineering*. New York: McGraw-Hill, 1974.

Osborn, A. *Applied Imagination*. New York: Charles Scribner's Sons, 1957.

Osgood, C.C. *Fatigue Design*. New York: Wiley-Interscience, 1970.

Pahl, G. and Beitz, W. *Engineering Design* (English edition translated by K. Wallace). London: The Design Council, 1984.

Palmgren, A. *Ball and Roller Bearing Engineering* (Third edn). Philadelphia: SKF Industries, 1959.

Patrick, R.L. (ed.) *Treatise on Adhesion and Adhesives*. London: Edward Arnold, 1967.

Peterson, R.E. *Stress Concentration Factors*. New York: Wiley-Interscience, 1974.

Pouliquen, L.Y. *Risk Analysis in Project Appraisal*. Baltimore: Johns Hopkins Press, 1970.

Pugsley, A. *The Safety of Structures*. London: Edward Arnold, 1966.

____ *The Theory of Suspension Bridges*. London: Edward Arnold, 1968.

Ray, W.S. "Complex Tasks in Human Problem Solving Research", *Psychological Bulletin*, Vol. 52, 1955, pp.134-49.

Reitman, W.R. "Heuristic Decision Procedures, Open Constraints, and the Structure of Ill-Defined Problems", Chapter 15 in M.W. Shelley and G.L. Bryan (eds), *Human Judgments and Optimality*. New York: Wiley, 1964.

____ *Cognition and Thought—An Information Processing Approach*. New York: Wiley, 1965.

Roark, R.J. and Young, W.C. *Formulas for Stress and Strain* (Fifth edn). New York: McGraw-Hill, 1976.

Rolt, L.T.C. *Isambard Kingdom Brunel—A Biography*. London: Methuen, 1957.

Rossmanith, H.P. (ed.) *Structural Failure, Product Liability and Technical Insurance*. Amsterdam: North Holland, 1984.

Rudd, D.F. and Watson, C.C. *Strategy of Process Engineering*. New York: Wiley, 1968.

Samonov, C. "Stiffness Constants in Bolted Connections", *Transactions, Institution of Engineers, Australia*, Vol. MC2, 1966, pp. 35-42.

Sandor, G.N. "The Seven Stages of Engineering Design", *Mechanical Engineering*, Vol. 86, No. 4, 1964, pp. 21-5.

Schon, D.A. *Invention and the Evolution of Ideas*. London: Social Science Paperbacks, 1967.

Serle, G. *John Monash—A Biography*. Melbourne: Melbourne University Press, 1982.

Shannon, C.E. and Weaver, W. *The Mathematical Theory of Communication*. Urbana, Ill.: University of Illinois Press, 1949.

Shigley, J.E. *Mechanical Engineering Design* (Fourth edn). New York : McGraw-Hill, 1986.

Smiles, S. *Lives of the Engineers - Boulton and Watt.* London: John Murray, 1904.

Sopwith, D.G. "The Distribution of Load in Screw Threads", *Proceedings, Institution of Mechanical Engineers,* Vol. 159, 1948, pp. 373-83.

Stark, R.M. and Nicholls, R.L. *Mathematical Foundations for Design—Civil Engineering Systems.* New York: McGraw-Hill, 1972.

Thompson, W.D'A. *On Growth and Form* (Abridged version edited by J.T. Bonner). Cambridge, U.K.: Cambridge University Press, 1961.

van Vlack, L.H. *Elements of Materials Science and Engineering.* Reading, Mass.: Addison Wesley, 1985.

Vickers, G. *The Art of Judgment.* London: Chapman and Hall, 1965.

Wallas, G. *The Art of Thought.* London: Watts, 1926.

Wankel, F. *Rotary Piston Machines.* London: Iliffe, 1965.

Watson, H.C. and Milkins, E.E. *Alternative Technologies in Motor Vehicle Fuel Conservation.* Report No. T48, Department of Mechanical Engineering, University of Melbourne, 1981.

Wehrli, R. *Open-ended Problem Solving in Design.* Unpublished Ph.D. Thesis, University of Utah, 1968.

Welford, A.T. *Fundamentals of Skill.* London: Methuen University Paperback, 1971.

Wen, Y.K. (ed.) *Probabilistic Mechanics and Structural Reliability.* New York: American Society of Civil Engineers, 1984.

Whittle, F. *Jet.* London: Frederick Muller, 1953.

Wilson, A.B.K. "The Powered Limb and Component Network—A General Theory Concerning Their Design", *Proceedings, Institution of Mechanical Engineers,* Vol. 183 (3J), 1968-69, pp. 7-10.

Woodson, W.E. and Conover, D.W. *Human Engineering Guide for Equipment Designers.* Berkeley: University of California Press, 1964.

Index

	Page	Section
abstraction	11, 16	1.3.2
accidents	1.3	
stimulus to design	5, 6	
theory	77-8	6.5.1 , 1.5.2
adhesives		13.8
aesthetics	20, 76	6.3.3
allowable stress		8.1.1
aluminum		9.2 , 9.4 , 9.5
A/M diagram		10.1.2 , 10.1.3
analogy(-ies)	33	2.3.3
anthropometry	73	6.2.1
anthropometric data	84-7	6.8
associative thinking	29,30	2.3.3
attitude to problem solving	34	2.3.3
automobile side pillar	53	3.5.1
beams		11.4
design		11.4
benefit	43	3.3.2
benefit cost analysis	44	3.3.2
bolted joints		13.1
design		13.1
fatigue		13.1
theory		13.1
tightening torque		13.1
bolted joint for jack hammer		13.9.1
brainstorming	30, 36	2.3.3
brass		9.4 , 9.5
bronze		9.4
buckling of columns		8.2.1 , 11.2.1
bug list	37	2.5.2
capital recovery factor	68	5.3
cast iron		9.2 , 9.3 , 9.5
Cayley	29	2.3.2
centrifuge		12.5
client	18	2.1
columns		11.2.1
column stress		11.2.1
design		11.2.1
theory		11.2.1
combinatorial search	31, 32	2.3.3
communications in design	11, 58-64	1.3.2
composite materials		9.5
compression members		11.2.1
compromise	52-56	3.5.1
conceptual blocks	34	2.3.3

concrete		9.6
condensing steam engine	28	2.3.2
constraint	21-3	2.2.3
contact stress		14.1 , 14.2
contingency factor	88	7.1
copper		9.4 , 9.5
cost	20	2.2.2, 3.3.2
automobile technology	69	5.4
expected total cost		8.4
life cycle	68-70	5.4
manufacturing	65, 90	5.1, 7.2
opportunity	68	5.3
creativity	27	2.3.2
flexibility	28	2.3.2
fluency	28	2.3.2
patterns of creative thinking	30	2.3.2
criterion	10, 17, 21, 23, 43, 45, 48, 52, 54, 55	
decision making	10, 41	1.3.2
sequences	45-51	3.4.3
sequence diagram	49	3.4.3
tree	47	3.4.2
deflection		8.2.1, 11.1, 11.4
degrees of freedom	45, 52	3.4.1
depreciation	69	5.4
design		
as engineering problem solving	1, 2	1.1
design space	45	3.4.1
design variables	45, 48-50, 52	3.4.1, 3.4.3, 3.4.4
flow chart of design process	12	1.3.3
hierarchy	13	1.4
method (IDEAS)		8.5.2
operational model	10	1.3.2
strategy	24, 26	2.2
dimension loop	90	7.2
dissociative thinking	29, 32	2.3.2
drawing desk	31	2.3.3
economic evaluation of projects	67	5.3
Edison	34	2.3.3
effective length		11.2.1
endurance limit		10.1.1
environment	20, 56	3.5.2, 6.4
ergonomics	71	6
evaluation of proposals	10	1.3.2
evolution of design problems	4	1.3.1
evolutionary design	44, 47	3.4.3
Euler equation		11.2.1
factor of safety	88	8.4.1, 8.4.2, 8.4.3
failure		
design against		8.1.1
design for		8.1.2
modes		8.2.1

fail-safe design		10.1.1
fatigue		8.2.1
design		10
life		10.1.1, 10.1.2
loads		10.1.1
feasibility	42, 44, 46	3.2, 3.4.2
feedback	15, 16, 51	1.5.1, 3.4.3
fillet welds		13.7.2
bending		13.7.3
bending and shear		13.7.5
direct loads		13.7.2, 13.7.4
fluctuating stress		10.1.1, 10.1.3
fracture		8.2.1, 11.1
full wave rectifier	94	7.6
function	27, 30, 76, 89	2.3.1, 2.3.3, 6.3.3, 7.2
functional dimension	90	7.2
gear design		14.5
gearbox, problems in design of	47-50, 96-8	3.4.3, 7.6
goals, in problem solving	3, 7, 14, 19	1.3.1, 1.5.1, 2.2.1
gripper for tensile tests		8.6.1
heat exchanger	53	3.5.1
idea log	36, 37	2.5.2
illumination	30	2.3.2
incubation	30	2.3.2
information acquisition	4	1.2
information display	74,75	6.2.4
information flowgraph	51	3.4.4
information load		
on designer	42, 45	3.3.4, 3.4.1
operator	74	6.2.3
information processing	11, 74	1.3.3, 6.2.4
initial appreciation	24	2.2.6
innovative design	44, 45	3.4.2
input-output analysis	22, 23	2.2.4
invention	28	2.3.2
jack hammer		10.3.1
jet engine	29	2.3.2
job satisfaction	76	6.3.2
Johnson parabola		11.2.2
laboratory chimney		8.6.2
leaf spring		9.6.1
materials		9.6.1
level crossing	4-7	1
load factor		8.1.1
loads on components		8.4.1
log-normal distribution		10.1.2
machinability		9.3
machine shaft		11.6.2
machining allowance		9.3
magnesium		9.2, 9.5

manned flight	29	2.3.2
manufacturing processes	93	7.4
characteristics	94	7.5
matching	24, 88	2.2.5, 7.1
material selection		8.5.2, 9.2, 9.7, 11
maximum principal stress theory		8.3
maximum shear strain energy theory		8.3
maximum shear stress theory		8.3
metaphor	33	2.3.3
models	43	3.3.2
mathematical	52	3.4.4
morphological analysis	31	2.3.3
motivation	76	6.3.2
networks	45	3.4.1
Newcomen	28	2.3.2
nickel		9.4
nickel-base alloys		9.4
notch sensitivity		10.1.2
objective	10, 17, 19-22, 42-4	1.3, 1.6, 2.2
performance	10, 20, 23, 93	2.2.2, 2.2.4
photocopier	22	2.2.4
Pilkington	34	2.3.3
pinned joints		13.4
design		13.4
theory		13.4
planning	3	1.2
plastics		9.2, 9.5
porridge words	16, 30	1.5.1, 2.3.2
postponement of judgment	36	2.3.3
power output, human operator	73	6.2.2
precedents in design		8.5.1
present worth	67	5.3
present worth factor	68	5.3
pressure vessels		12.1
classification		12.1
design		12.2
probability	43	8.4.4
problem	2, 3	1.2
definition	10, 19	1.3.2, 2.2.1
exploration	10	1.3.2
recognition	10, 30	1.3.2
problem solving	1, 2	1.1
process variables	4	1. 2
task variables	4	1.2
product design	19	2.2.2, 2.2.4
production-consumption cycle	20	2.2.2
professional engineering behaviour	1, 2	1.1
psychological set	34	2.3.3
rate of return	66	5.2
reliability	55	3.5.2
report writing	59	4

resource allocation	3	1.2
reversed stress		10.1.1, 10.1.2
risk	43, 91	3.3.2, 7.3
rolling element bearings		14.1, 14.5
safe-life design		10.1.1
safety	77	6.5
cost of safety	78	6.5.1
saturation	30	2.3.2
search for proposals	10	1.3.2
sensitivity to problems	18	2.1
shafts		11.5
design		11.5
theory		11.5
shear connections		13.5
signal detection	73	6.2.3
simulation, Monte Carlo	43, 92	3.3.2, 7.3
size effect		10.1.2
slenderness ratio		11.2.1
S/N diagram		10.1.1
solution specification	10	1.3.2
spur gear drive	47-9	3.4.3
stability		
structural		8.2.1, 8.2.2
system	77	6.5.1
Standards	92-3	6.2.1
steel		9.2
alloy		9.3, 9.5
carbon		9.3 , 9.5
high strength low alloy		9.3
stainless		9.3, 9.5
tool		9.3
steel making		9.3
steels		
heat treatment		9.3
tree of		9.3
stereotypes	75	6.3.1
strategy		
computational	51	3.4.4
decision making	45	3.4
design	24, 26	2.2.6
graph	53	3.5.1
strength of component		8.4.1
stress concentration		8.4.3, 10.2, 11.5
stress concentration factor		10.1.2, 10.2
stress factor		8.1.1
structural integrity		8.1.1
student work area	79-83	6.7
surface fatigue		14.4
surface finish		10.1.2
synectics	36	1.5.1
synthesis	15	1.5.1
systems	13, 21, 71-2	1.4, 2.2.2, 6.1
associative	24	2.2.5
flow	24	2.2.5
tangent modulus		11.2.2

task variables, in problem solving	4	1.2
theory of failure		8.3 (4)
thought experiment	15	1.5.1
timber		9.6
titanium		9.5
tolerances	89	7.2
dimensional	90	7.2
performance	95	7.6
toothpaste dispenser	37-40	2.5.2
tracking	75	6.2.5
trade-off	55	3.5.2
transformation, problem data and language	34	2.3.3
trial and error	15	8.5.2
trouble shooting	3	1.2
truss		11.6.1
uncertainty	88	7.1
in design for integrity		8.4.3
in design variables	91	7.3
values	4, 43	1.2
vigilance task	74	6.2.4
water heater	7-9	1.3.1
Watt	28	2.3.2
wear		14.4
Weibull distribution		10.1.2
welded joints		13.7
see also fillet welds		
wheelchair	24-6	2.2.6
Whittle	28	2.3.2
worst case analysis	90	7.2
yielding		8.2.1 , 11.1